| Introduction | 1 |

The 21st Century View of Human Nature and Reality ... 6
THE SUCCESS OF REDUCTIONISM IN SCIENCE ... 6
PROBLEMS WITH REDUCTIONISM ... 13
The Problem of Complexity and Emergence ... 14
The Greatest Complexity - Life ... 16
Chaos and Complexity ... 21
THE CHANGING PARADIGM ... 26
The First Revolution: Quantum Mechanics ... 28
Quantum Weirdness ... 34
The Second Revolution: Einstein's Thoughts on Relativity ... 42
"Theories of Everything" ... 46
Summary 58

The failure of Artificial Intelligence. ... 61
Chess and computers ... 62
'Go' and computers ... 66
Language and Computers ... 68
The Challenges to Artificial Intelligence ... 72

Subjective consciousness ... 77
Penrose, Chalmers and the Brain/Mind Debate ... 78
Philosophical Schools of Thought on Consciousness ... 82

The great Humpty-Dumpty disaster in Neuroscience ... 85
The Grandmother Cell and Top Down Approaches ... 85
The 'Binding Problem' ... 87
The Being Within – the Self, 'Homunculus', or the 'Cartesian Theatre' ... 89
CRITIQUE OF THE BEHAVIORIST OR 'PHYSICALIST' VIEW – FOR WHOM IS THE ILLUSION? ... 97
CONFUSION AMONGST THE PHYSICALISTS ... 102

The Arrow of Time and Subjective Consciousness ... 110
The subjective sense of Time ... 111
Breathing fire into the Equations ... 117

Free Will, the Self and the sense of the numinous – not just Illusions! ... 119
Evolution of the Human Brain, Mind and Will Power ... 119
The 'God centre' of the Brain, or the 'Brain centre' connection to an 'extramundane reality'? ... 126

Genome of a few Megabytes – the Emperor's new Genome ... 132
The Structure of the Genome ... 133
The old version of Darwinism and the 'selfish gene' ... 137
The missing genes ... 145

Operating System, Computer Language and Code all in the Genome? 157
The Complexity of the Body / Brain system 160
EXAMPLE OF COMPLEX DETAIL: EXTRACELLULAR MATRIX SECRETION CELLS 167
Fractal compression in the Genome, or algorithmic complexity? 170
Socio-biology, Evolutionary Psychology or a new form of 'Vitalism'? 173

The Emperor's new Meme 179

More things in Heaven and Earth 182

Conclusions: The Meaning of life, the Universe & Everything 206

References 209

Introduction

This book explores three related areas which lie at the cutting edge of modern science and which therefore all share the exciting prospective of future discoveries. The three areas are (physics / complexity theory), genetics, and consciousness studies. All these areas are crucial to our understanding of ourselves and our place in the universe. All three bear on the question of human nature and thus, in a sense, the ' meaning of life'. Genetics has to do with how our bodies arise in a seemingly miraculous way from a single cell. Consciousness is the mysterious aspect of our inner lives which seems to breathe something extra into the body whose parts, the proteins, are determined by genetics and yet which is much more than the sum of its parts. To understand the interplay of genetics and consciousness in the context of a living body, we will have to deal with, among other things, complexity theory and its links to the weird world of quantum physics.

The tendency in the media nowadays is to report on findings in these key issues in science as if they were foregone conclusions. Yet in each of them the future is wide open, offering untold possibilities. Take for instance the consciousness research discipline of neuroscience. There is a bias for only those findings which imply that our brains are nothing more than souped-up computers to make the headlines. Seldom is anything mentioned of other trends in mind-brain research which point to the still mysterious way in which our conscious experience unifies the input of the senses to generate the internal television screen. The books which are reviewed in the major papers tend to be from neuroscientists reporting on what they have learned from cases of brain-damaged patients. And if a philosopher of consciousness is mentioned it is almost certain to be Daniel Dennett. The impression left from these sources is that the brain has been completely mapped in terms of function, and that consciousness, ego and self are all illusions. In this book I will point out how, contrary to the picture painted in the media, much is still unknown and mysterious when it comes to consciousness and the interaction of mind, brain body, genes, biology, physics & chemistry. That's a heady mixture, but a fascinating one.

I will try to emphasize the more holistic view of our mind-brains, and why they are a lot more than the sum of their parts. I will also emphasize the positive aspects of our brain-mind systems, which the disease-oriented researchers and nay-sayers so beloved of the media often fail to mention. The immense power and staggering complexity of our brain-minds functions so efficiently and magnificently when they are healthy that they have become an embarrassment to the ultra-materialists and heroes of the Artificial Intelligence scene. The boast of the latter, made repeatedly over the past decades, that a conscious computer was only a few years away, is as remote from fulfillment as ever.

Another area that receives one-sided attention is genetics. Some years ago, it was thought that the human genome should contain hundreds of thousands of genes, in order to explain the complex structure of the human brain and body. However, now we know that this is not the case, since the shock-horror result of the genome analysis showed we have only between 20,000 and 23,000 genes. In this book I explain some of the consequences of this situation for our ideas about human nature. One of the steps in this explanation involves estimating the information content or complexity of the genome. I have tried to keep this explanation clear and simple, as in previous essays on the topic I was surprised at the incomprehension that met what I thought were straightforward calculations. For the latter reason I will proceed more slowly here, in the hope that the ideas will appear as self-evident to the reader as they do to me. I believe a book dealing with this topic is needed, as in several popular science books recently, as well as in articles in Scientific American and the like, there was a surprising tendency to avert the eyes from the central truth of the matter. It is as if a mental block of some sort prevents most science writers from grasping the nettle of the 'missing genes'. This missing complexity in the genes is all the more amazing when we consider the incredible complexity of the brain and consciousness. And recent research in epi-genetics as of 2010 does little to solve the mystery – the extra information gained from some RNA sequences and gene suppression does not increase the information content significantly.

Though this book seemed more than ever necessary in the aftermath of the first version of the decoding of the human genome in 2000, the germ of the idea was sown many years before. The first awakening of this germ was my study of quantum mechanics. The second stirring of doubt in the prevailing paradigm came with my reading of that critique of Artificial Intelligence, "The Emperor's New Mind" in the mid 90's. After the second, more detailed genome breakdown published in 2004, when the number of genes shrank further from 30,000 to nearer 20,000, these doubts had grown to a great fat question mark.

The risk of writing against the prevailing opinion on genetic influence, evolution and the brain/mind is that one will be branded a creationist. Thus let me at this point make it clear that I will have no truck with the creationists. I am also not a new-age type esoteric writer. My brand of mystery is based more on the wonders found mostly within science and philosophy, and much less in the woolly thinking of the new age fraternity. With this in mind, the plan of the current book took form. I start by examining interesting questions in 3 areas:

- Physics - Quantum Mechanics showed that matter is not composed of 'Newtonian billiard balls'. What, then, is matter?
- Psychology / Philosophy / Neuroscience - Are we really just machines?
- Genetics - is there enough information in the

genome to specify 'the machine' implied by the simple view of Darwinism?

This examination leads on to a discussion on the nature of the current paradigm and its failings. I hope that it will serve as a worthwhile contribution to the debate in this area. If it helps clarify the significance of the latest research for even a handful of people then it will have been worthwhile. Also, I would hope that it might stimulate some readers to think the matter through for themselves and thereby form their own unique point of view. That is very much in keeping with the tenor of the book, which stresses individuality over the collective, uniqueness over uniformity, and awareness over zombie-like somnolence.

The 21st Century View of Human Nature and Reality

The growing success of science during the past 500 years is an impressive story. Once the scientific method had been invented, it could be applied to many of the age-old problems. The difference compared to pre-renaissance science was that the new method was for the first time systematic and well ordered. By breaking down large problems into manageable smaller problems, the new method of practical experiments accompanied by theory building expanded knowledge in all fields of science, from physics to chemistry and biology.

The Success of Reductionism in Science

The careful dissecting of intractable problems into smaller ones became known as 'reductionism'. I.e. things were reduced to their simplest components. In this way, complex problems could be broken down into a set of smaller, simpler ones more amenable to solution. Thus matter was analyzed on ever smaller scales until physicists confirmed its atomic nature. So the old thought experiments of the Indian philosopher Kanada (6th century B.C.) and of the Greek Democritus (5th century B.C.), which supposed that matter was composed of fundamental building blocks called atoms, were confirmed by the new findings in chemistry and physics.

In the 20th century it became apparent that the atoms were not the smallest constituents of matter, as they were in turn composed of the sub-atomic particles called protons, neutrons and electrons. It is now accepted that protons and neutrons have further sub-structure, being composed of three 'quarks'. String theory and other theories such as Quantum Gravity propose that there are much smaller components of these sub-atomic particles such as 'strings'. The latter are of the order of the Planck length, which is about 10^-

^{35} m, while a typical atom is estimated to be approximately 10^{-10} m. The size of a proton or neutron is about 10^{-15} m. As an example of an atom in terms of its sub-atomic components, see Figure 1 below, where I show the quarks inside the 2 protons and 2 neutrons of a Helium atom. The quarks involved here are the Up quark (U) and the Down (D) quark.

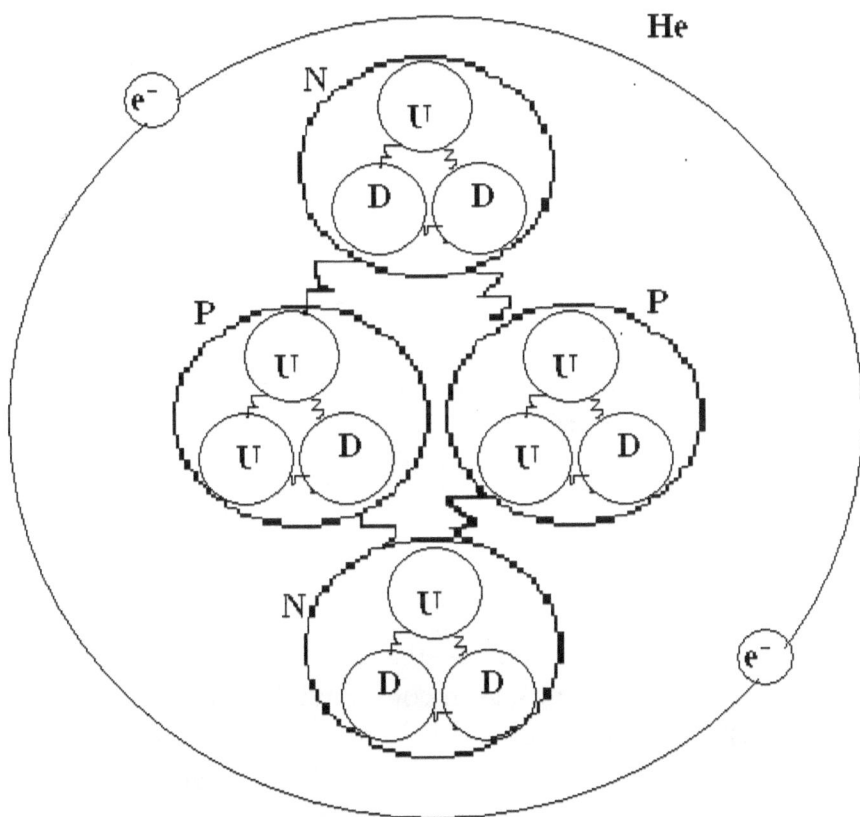

Figure 1 A Helium atom showing the quarks inside the protons)

(P) and Neutrons (N). (Drawing by author).

The effect of quantum uncertainty at the sub-atomic level make it extremely unlikely that structures smaller than the Planck length of 10^{-35} Metres will be discovered. But the picture is already clear. The reductionist method in physics has been extremely successful. The same may be said of chemistry. Different materials were broken down so that it could be shown that everything was a mix of basic substances called 'compounds'. The compounds were found to be composed of 'molecules', or clusters of different types of 'atoms'. These different atoms were found to be held together by different sorts of bonds. If they shared electrons the bonds are called covalent. These electrons are shared in order to complete a stable shell of electrons – usually 8. Thus as carbon (C) has 4 electrons in its outer shell, and oxygen (O) has 6, a C atom can share 2 electrons each with two O atoms to give a linear molecule (O = C= O) or CO_2, the infamous greenhouse gas of global warming.

Other molecules are held together by different types of bonds, e.g. ionic bonds, as in common table salt, Sodium Chloride. In ionic bonds, some atoms capture extra electrons, giving them a net negative charge, while others lose some electrons to gain a net positive charge. These positively or negatively changed atoms are called ions, and it is the attraction between the positive and negative ions that hold ionic substances together. Many insights into the structure of crystals, organic (involving carbon) and inorganic molecules have been obtained by modern chemists. The result has been an explosion in the number of novel substances being artificially created — some famous examples being nylon, non-stick pans and plastics. Soon chemistry may revolutionize production of solar cells — with cheap chemicals replacing expensive crystalline Silicon.

Also in biology, great strides were made by following the program of reductionism. Thus with more powerful microscopes it was possible to show that all animals and plants were made up of tiny 'cells'. Each cell is similar to the smallest independent animals, namely, bacteria or protozoa. Just as the atom was shown to have sub-structure, so were cells shown to have internal sub-structure. While in bacteria, there are free-floating strands of DNA and relatively simple processes (though still complex compared to any machine), in protozoa and animal and plant cells the structure is far more complex. There are several sub-cells or bubbles within the main cell, referred to as 'organelles'. One of these organelles, if a rather special one, is the nucleus. In Figure 2 below, this sub-structure is indicated in a very schematic way. The nucleus is shown, which contains the Genome or most of the DNA which controls how the various proteins are built. The DNA is now protected from attack (e.g. by viruses), unlike the exposed DNA of bacteria. Also shown explicitly are Ribosomes, which interpret messenger RNA from the Nucleus to build the proteins, and so may be thought of as protein factories. The Mitochondria are more like power stations, as their task is to convert organic molecules into an easily usable form for use in the various reactions that control the cell's metabolism.

In fact the processes going on in even the simplest cell are extremely complex. It would take many chapters to describe in any detail the processes of a simple bacteria like e-coli. For the next level of complexity after bacteria, namely the protozoa like Paramecium or Amoeba, the structure is of higher level of complexity, as shown below in Figure 2, and what's more, the processes are more directed, involving complex interactions and cascades of different types of protein. Such cascades have been painstakingly worked out by microbiologists over the last 30 years. Initial hopes of finding a simple one-to-one relation between genes, proteins and functions were dashed as the incredible complexity revealed itself.

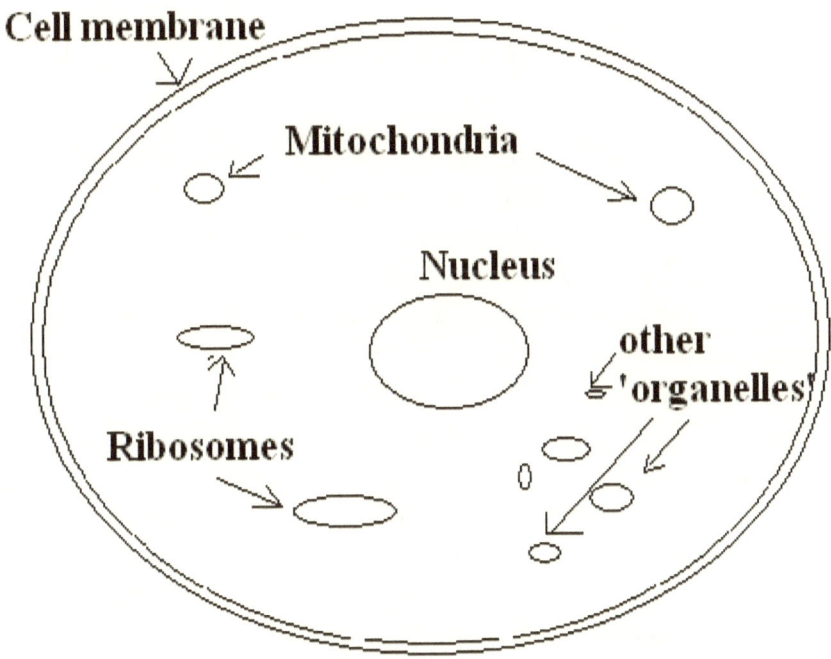

Figure 2: A living cell, showing some of the 'organelles' (see text). (Drawing by author).

We now know that characteristics of animals or plants are in general not a simple function of an isolated gene. In the vast majority of cases, multiple interactions of genes, proteins, RNA etc. are involved. This explains why the initial promise of genetic therapy in medicine was not met. Only a few very rare diseases had such a simple cause, making them amenable to simple genetic treatment.

Problems with Reductionism

As indicated in the discussion of the cell above, the simple view of one gene creating one protein, with a problem in this process causing a disease, was found in general to be incorrect. Another indication that straight-forward reductionism would not always work was the discovery that chaotic processes were important in nature. As indicated by Margulis and Sagan (1995), Capra (1996) and Sheldrake (1981), the development process for any animal or plant is characterized by many feedback loops, with complex interactions between proteins and the genome. This sort of interaction is not 'linear' (e.g. a leads to b which leads to c) but highly non-linear (e.g. a with b and w lead to d, which with y, q and r leads to ca and gz, which together with f leads to a). Also, each new level of complexity involves new 'emergent' properties, i.e. properties which imply that the whole is more than the sum of the parts.

The Problem of Complexity and Emergence

Physics has equations for dealing with simplified cases – e.g. Newton's law of gravity may be applied to any 2 bodies such as planets floating in space. However, if a third body of similar size enters the scene, the equations may no longer be solved exactly. A computer is then needed to calculate ever better approximations to a solution.

Thus Newton's law only caters for cases of the utmost simplicity. For 3 or more bodies in general the computer solution must be used. This is known as the 3-body problem. This already gives an idea of the difficulty that science has in dealing with complexity. Another case is on the atomic level. Exact solutions of the quantum mechanical equations for atoms and sub-atomic particles (Schrödinger's equation) can only be obtained easily for the simplest atom, hydrogen, which consists of just one proton and one electron (see figure 1). To solve exactly for complex molecules is impossible. This is known as the many-body problem in quantum mechanics. Again computers must be used to get an approximation, and for any reasonably large system the amount of computer time needed to obtain an approximate solution is large.

In general, for any large number of molecules it is impossible to model exactly all the interactions. Even approximations of the individual interactions become too complex. One must then resort to statistics, as in the equations for an ideal gas. Concepts such as pressure are average measures of the impact of many billions of molecules on each square centimeter. Another way of looking at it is that billions and billions of atoms moving freely have properties which gradually emerge, i.e. the properties of a gas.

Another example, not quite as dramatic, is a chess game. In principle it could be so simple: just 6 different types of player (rook, bishop, knight, king, queen and pawn) moving on a very limited board of 64 squares. Yet very soon complex and unique situations arise that only the best chess players or most powerful computers can solve.

On the same theme, an alphabet of 26 or fewer letters is enough to generate all the works of literature, science etc. that cram the bookshelves of the world and jam the internet. In a way, a computer can encode everything using just 1 or 0.

Another aspect of complexity is the emergence of new and 'unexpected' properties in combination of simpler systems. An example of this is the liquid characteristics of water – these are not easily explained as a consequence of the properties of Hydrogen on its own or Oxygen on its own. Burn Hydrogen in Oxygen and the result, water, is something new and with new properties. I.e. the whole is more than the sum of the parts. Yet another example from chemistry is ordinary table salt: take the highly reactive metallic substance

Sodium and combine it with the poisonous gas Chlorine and the result is the harmless, and even edible, salt or sodium chloride.

Other cases of emergence are the bodies of animals or plants, which are just many single cells working together. But out of that cooperation a new phenomenon emerges, namely the large (macroscopic as opposed to microscopic) animal or plant with all its organs and abilities. This reaches a pinnacle in the human, where the complexity of the body, already great, is accompanied by the complexity of the human brain. Our brains are said to be the most complex systems in the universe. This may be true, unless there are aliens with even more powerful brains.

The Greatest Complexity - Life

This is one of the main characteristics of complexity, then. Unexpected properties may emerge when the complexity increases – e.g. when the number of particles or cells involved exceeds a certain limit or two different chemicals interact.

Another example of emergence is on the global scale: cities emerge from many people interacting economically and culturally, and countries emerge from cities and their interaction with the countryside. Ecosystems emerge from many animals, plants, fungi, atmosphere and water interacting. The entire planetary system, the ultimate example of an eco-system, often seems to be a self-regulating whole. James Lovelock (1979) in his Gaia hypothesis, implied that the climate of Earth is maintained at a temperature that is comfortable for living things. He was led to his hypothesis when, as a member of the NASA team planning the first experiments to measure life on Mars, he asked himself what to look for.

What made the Earth different from other planets, Lovelock realized, was the unusual combination of gasses in the atmosphere. Thus an alien astronomer with a powerful spectroscope for measuring the composition of planetary atmosphere would see that the highly reactive and unstable gas Oxygen was present in the unbelievably high concentration of about 20%. The unusual mix of nitrogen, oxygen and carbon dioxide would be recognized as the chemical signature of some 'unnatural' process, since basic chemistry would have achieved an equilibrium much closer to the atmosphere of Venus. Venus, our sister planet, is almost exactly the same size as Earth but instead of our pleasant garden paradise it has a burning hell of raging wind and an atmosphere almost entirely of carbon dioxide. The latter gas, as we know from the fears of global warming on Earth, is an ideal greenhouse gas, which accounts for the high temperature of the surface of Venus (over 400 C).

So our alien astronomer would not have to send a probe to Earth to conclude that life was present.

Lovelock's insight was that life has to create its own environment in order to survive. There must be a global supportive infrastructure to maintain ideal conditions for the non-equilibrium processes that characterize life. Lovelock further concluded that life is responsible for several chemical cycles that prevent the build-up of harmful gasses. One of these involves the movement of tectonic plates, which through subduction of crust material buries in the Earth's mantle much of the carbon that would combine with the oxygen to give the greenhouse carbon dioxide. By this sinking of the crust in the lower mantle layers, the skeletons and remains of bacteria and other life forms, rich in carbon, are buried for millions of years. Note again that this is another difference of Earth compared with its sister Venus – i.e. the tectonic plate movements. The surface of Venus is one giant tectonic plate. No circulation occurs via plate movement, which not only stops materials being sub-ducted, but also removes a major way to get rid of heat building up in the planet's interior. So for the internal heat of the planet to escape on Venus, large shield volcanoes have to form. The surface of Venus is littered with volcanoes, many of them of the large shield variety akin to that of Olympus Mons on Mars. On Earth, by contrast, the circulation of the tectonic plates in sinking at subduction zones and rising in other places like the mid-Atlantic ridge, allows heat to rise to the surface gradually without catastrophic build-ups as on Venus. Another aspect of this is that the sea beds on Earth allow the continental plates to move around. This ease of plate movement may be partly due to our unusual moon, which is now widely believed to consist of material from the primeval Earth's outer layers, scooped up by the impact with a large body in that violent era of the early solar system. Thus the moon has hardly any iron, as it is thought that Earth's iron had already sunk to the core when this impact occurred, so that the outer non-metallic layers became the main constituents of the moon. So most of the crust was lost to the moon. The result is that the remaining crust can move around in the ocean basins. On Venus, where this collision didn't happen (though some collision did occur, which might explain its strangely slow rotation about its axis), the crust is 'grid-locked' and cannot move around as on Earth.

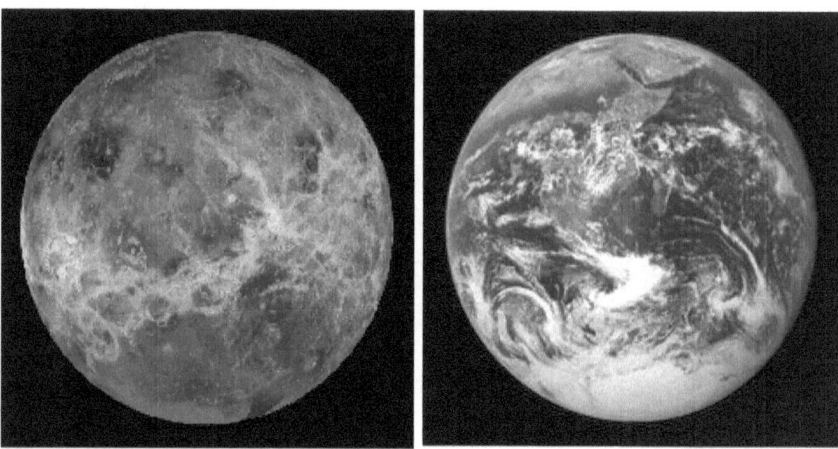

Figure 3: Unlike sisters: Venus (left) is an arid hothouse planet, while Earth (right) is a pleasant garden planet with oceans and mild temperatures suitable for life (pictures courtesy of NASA, via Wikipedia).

So the presence of our moon has many advantages. Apart from perhaps having allowed tectonic plate movement, it also stabilizes the rotation of the Earth, so that it doesn't tumble around, as may have happened with Venus, since the latter is the only planet to rotate clockwise, looking down from 'above' our north pole, implying that it may have slowly tumbled into this orientation. Another extreme example of this sort is Uranus, whose axis of rotation lies at 90 degrees to the average north-south direction in the solar system. Such an unstable rotation axis would cause major climate changes, which would be difficult for evolution of any complex life forms. The tilt of a planet's axis of rotation with respect to its orbital plane is known as its 'axial tilt' or 'obliquity'. Research has shown that the obliquity of the other planets appears to vary chaotically over the millennia, while the Earth's remains relatively stable (see e.g. Neron de Surgy & Laskar 1996), again thanks to our moon. This effect is one of many indications that Earth is a very privileged world.

It is always amazing to contemplate that whilst planets like Venus and Mars are relatively simple, the Earth harbors the most complex systems in the solar system and maybe in the galaxy, if not the universe.

Before the discovery of genes and proteins and the connection between them, the complexity of life appeared to need some mysterious force. This force was referred to as 'elan vital' or life force. Since the discovery of genes, it was thought that all life forms could now be explained as mechanisms controlled by the genes. Thus opinion swept from one extreme to the other. From a time when every living thing had a soul to a time when all creatures were merely mechanisms of a deterministic, materialist nature. These mechanisms could be reduced to their smallest components, the genes, molecules and atoms. This reductionism neglected the phenomenon of emergence, however, whereby the whole is more than the sum of its parts. This factor will be discussed in greater depth later in the book.

Chaos and Complexity

Another aspect of complexity is the phenomenon that has come to be known as chaos by scientists. In the sense of 'chaos theory', this often involves equations sensitive to initial conditions. A famous analogy is the 'butterfly effect', which says that the equations governing the weather on our planet are so complex and non-linear that even a butterfly flapping its wings in Beijing can cause a disturbance which grows in unpredictable ways, maybe even resulting in a storm in Europe.

There are basically two sorts of chaos – classical and quantum. In the classical case, the particles are sometimes at least in principle deterministic, i.e. if we knew the starting conditions well enough we could predict exactly how all particles would behave in the future. However, even for a relatively simple cases such as the motions of solar system bodies, this is not strictly the case (recent development in planetary theory imply that the Solar System is weakly chaotic, but over a very long time. Smaller planets like Mercury and also the asteroids are more chaotic in their orbits). In a system such as the atmosphere, with so many variable factors acting on countless billions of atoms, the task of knowing the initial state is an impossible dream. Well, maybe not impossible, but very, very impractical. In the quantum case, however, it is not even possible in principle to predict what will happen, as Heisenberg's uncertainty principle foils every attempt to know the state of atoms to unlimited accuracy.

The most famous example of dramatic images from chaos theory involve 'strange attractors', and fractals of certain mathematical systems. The best known fractal is the 'Mandelbrot set' – see Figure 4 below, where the pictures focus on greater detail we go left to right.

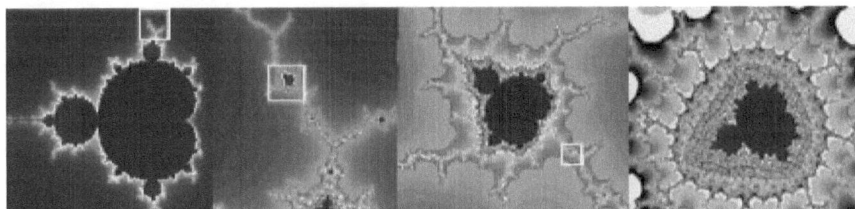

Figure 4 – Increasing magnification of sections of the original Mandelbrot Set pattern (left). The insets focus in, closer and closer. (picture taken from Wikipedia).

Note that on the images on the right we see copies of the original pattern. This is one of the amazing properties of the fractal – that it is 'self similar' at all scales. Yet all this complexity arose purely by iterating a simple equation. This is another example of classical chaos: by varying slightly the simple seed equation, the fractal takes on a radically different form. Fractals also show how great detail can be compressed into a very compact form. From the example of a fractal fern (see figure 5 below), we see how the similarity on all scales really does occur in nature.

Figure 5 – A 'sub-leaf' of a fern looks like a 'leaf', which looks like the fern itself. (image taken from Wikipedia).

Another example of fractals in nature is a coastline – e.g. that of Norway: no matter what scale you use – 100's of km, 10's of km or km, the ragged structure appears similar.

Fractal compression of information is used in some methods of compressing computer images. One question is whether the genome compresses information in a similar way, allowing vast amounts of complexity to be stored in a very compact form in the genome. We will return to this question in the chapter on the "Emperor's New Genome".

The Changing Paradigm

For most of the 20th century, because of the success of reductionism, it became normal to assume that anything could be understood by breaking it down or dissecting it. This in turn led to a "mechanistic-materialist" philosophy – i.e. that everything, including the human body-mind, was a mechanism or machine-like system that could be understood by looking more closely at its parts. This philosophy was the dominant one, and to a certain extent still is the default philosophy of mechanistic, reductionist science. This is so much the case that it is usually only present as an invisible, implicit assumption. It is seldom that it is explicitly stated in papers in different areas of science (genetics, chemistry, biology etc.) that this philosophical theory is being used. It is simply taken for granted. Yet it is just that, a theory, for which the proof is no longer all embracing, as we shall see.

We saw a hint of the deficiency of the old reductionist mentality in the discussion above of emergent phenomena that implied that the whole was more than the sum of its parts. In the case of the brain/mind system, there is reason to believe that something more is going on than "simply" emergence. It has also become apparent that a linear view of causation breaks down for systems like the brain: research in the 1990's showed for example that monkeys alter the response of the area of their retina showing something of interest like a banana. That's an example of feedback and nonlinearity. In the old linear view, the banana image is passively formed on the retina and processed upward to higher levels of processing and awareness. In this older view, the process was one-way, always from lower levels in the sensory hierarchy to higher ones, i.e. always bottom-up. But in the new view there is feedback from consciousness to the level of the retina, which is a top-down, non-linear process. I.e. the mind-brain must have a model of the banana and sent signals all the way back down though the network to enhance the signal. I will come back to this later in the discussion of consciousness.

First, let us look at the changing world view in physics over the last century. The revolution in physics started at the beginning of the 20th century. Both of the main revolutions, Einstein's Relativity theory and Quantum Theory, involved introducing the observer into the equation for the first time.

The First Revolution: Quantum Mechanics

Before the 20th century, Newton's view of the world was the dominant one in the increasingly industrialized and technologically oriented western society. The Newtonian view of atoms was as hard, ball-like particles, like miniature billiard balls. These billiard balls moved around under the influence of physical and chemical forces, like cogs in an immensely complicated piece of clockwork.

The first bombshell that implied that this Newtonian world-view was insufficient to describe reality came between 1895 and 1900, when Max Planck was forced to assume that electromagnetic energy (visible light or Infra-red, Ultra-violet, etc. radiation) was not continuous, but comes in discrete packets or quanta (called photons in the case of light), where each quantum of energy was an integer multiple of a basic constant, h, now called Planck's Constant. This conclusion was based on his analysis of data on the spectrum or color of heated bodies. Think of a horse shoe in a blacksmith's forge. As it gets hotter and hotter, it first glows red and then white hot. Its color or spectrum changes with its temperature. For experimental measurements physicists preferred heated cavities with an observation window for a spectrometer. These cavities also changed color or spectrum when heated, just as the horse shoe.

Another aspect of Planck's discovery was that light quanta or photons were sometimes like waves and sometimes like particles in their behavior. This shattered the proponents of the 'wave only camp' and the 'particle only camp'. For centuries that debate had raged. Huygens, for example, explained light's optical properties such as reflection, refraction etc. as being due to its wave-like nature, while Newton considered light to be a stream of particles. Both the wave and particle views had evidence to support them. Then with Planck, Einstein & co. Nature showed its ability to engage in a dialectic, combining thesis and antithesis in a synthesis that was the new wave-particle duality. How difficult it was to abandon the old either/or paradigm where it could be 'wave only' or 'particle only' was shown by Bohr and others hanging on to one or other of the 'alternatives' until the bitter end, when wave/particle duality was shown to be a reality beyond any shadow of a doubt.

Such a synthesis of alternative theories as the wave/particle nature of light had, before the 20th century, never been seen before in science. Up to then it was always the case that competing theories fought it out for years until one gathered so many points of experimental evidence that it was declared the winner. Thus most physicists reacted violently against the new quantum mechanics at first. Only when it was shown to be the 'most accurate theory ever' was it reluctantly accepted by the vast majority as a practical way of describing results. The philosophical implications were studiously ignored for many years, as most physicists were afraid of the abyss that might open there.

The process of only allowing those energy values that were a multiple of ' Planck' s Constant', or h (actually h.mu, where mu is the frequency of the type of light under consideration) was called quantisation. Planck was forced to 'quantise' energy in order to explain the observed spectrum of 'black body radiation'. 'Black Body' is a bit misleading here, as in fact the light has many different colours – only they may not be in the visible range of the spectrum (e.g. infra-red or ' heat radiation' can be felt but not seen). Classical physics, which assumes that light may be emitted from atoms in a continuous range of frequencies, gave an absurd prediction for the energy distribution, namely that the energy per unit frequency increased indefinitely with increasing frequency. This was called the "great ultra-violet disaster", as the classical solution broke down for UV light and light of higher frequency. See Figure 6, which illustrates the difference between the classical and the Planck distributions. It may be seen that Planck's solution reproduced the observed data much better at high frequencies, although the classical solution was also good for lower frequencies.

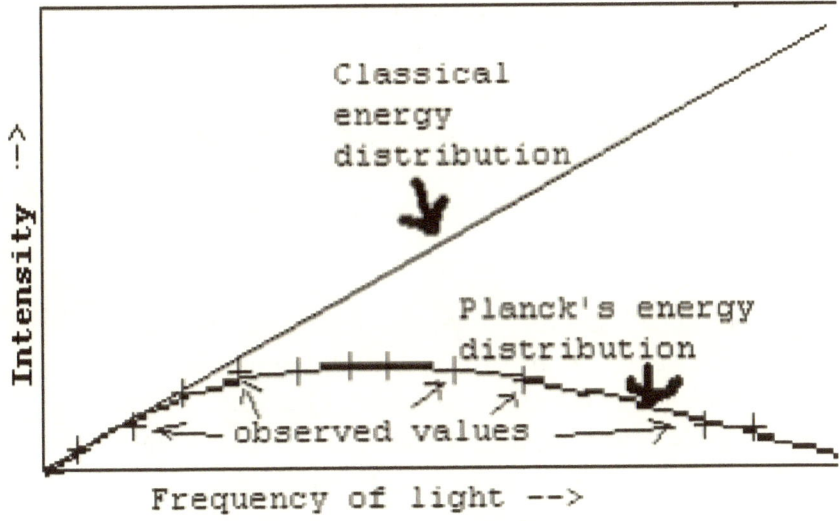

Figure 6: Comparison of classical and Planck's or quantum energy distributions (curved lines as shown) with observed values (crosses).

Planck's method avoided the "great ultra-violet disaster", but at the cost of quantising energy. The consequences of this were far-reaching, as once energy is quantised it turns out that many other quantities, such as momentum and position, are also quantised. As others developed the quantum theory further, it became apparent that even stranger consequences had to follow from Planck's original insight. Heisenberg, Schrödinger, De Broglie and others showed that one could not speak of a particle being in a definite position at a definite time. There was always an element of uncertainty to any measurement. This in turn showed that there was no such thing as matter in the classical sense of hard and fast billiard balls. Each particle, instead, was described by a wave called the 'wave function', usually denoted by the Greek letter Phi. This mysterious quantity has the property that when multiplied by itself it gives a probability density function. This just means that e.g. Phi . Phi . (x2 – x1) gives the probability that the particle may be found between two positions x1 and x2 in some direction.

Before the 20th century physicists had modeled matter as hard, concrete stuff (the exception being light, which was modeled as waves in the ether). Quantum mechanics, however, now showed that it was nothing of the sort - it was a web of ghostly probabilities, the 'wave functions' of 'particles' such as atoms, protons and electrons. We are literally such stuff as dreams are made of. Experiments toward the end of the 20th century showed that even objects as large as atoms, which can be photographed with electron or tunneling microscopes, can 'interfere' with themselves as if they were just ghost-like probabilities of being in two different positions at the same time (Horgan, 1996, Choi, 2003). To a physicist of the 19th century it would have sounded like fantasy to say that an atom could exhibit bi-location, walk through walls (quantum tunneling) and anticipate what might have been (quantum interference).

Another key point about the Quantum view of the world was that the observer suddenly entered the picture. Note that before Planck, science sought to distance itself from any implication of bias or subjectivity. Thus all science reports were, and to a large extent still are, phrased in a very cold, 'objective' style. For example: "A beaker was cleaned and 2 ml of water were poured into it. Etc." But the subjective aspect of things exists and is important and is not only characterized by bias or inaccuracy. This point was bound to come out sooner or later and in Quantum Theory it came out loud and clear.

An example of the importance of the observer in Quantum Theory is in measuring the properties of atoms or sub-atomic particles. An example of such a property is the spin of an electron, which is similar but not quite the same as the spin of a large object on its axis, such as a ball or a planet. Another such property is the direction of vibration or 'polarization' of a photon.

The physicist sets up an instrument to detect the spin or polarization of the particle. Let us take the case of spin. Before measurement, the spin is not well determined – it is thought to be in that characteristic grey quantum state where nothing is totally certain. When the measurement occurs, however, the spin is suddenly clearly defined. The instrument reads out a fairly accurate value. This is the characteristic crystallization of a quantum state, sometimes referred to as the collapse of the wave function, where the 'wave function' is the quantum 'grey state'. One of the key points here, though, is that just by choosing the direction of the instrument's measurement axis, the particle is forced to give a fixed number of quantized spin units along that axis. In the case where the electron spin is the lowest possible (which it is), the observer has by her act of measurement determined in which plane the spin is occurring.

Quantum Weirdness

Another observer-related effect in quantum mechanics is the famous case of 'Schrödinger's cat'. In this gruesome experiment, a cat is thought to be in one of those quantum grey states where nothing is certain. For the sake of the thought experiment, there is an even chance of the cat, sealed in an opaque box, being alive or dead after one minute. Schrödinger speculated that the cat was simultaneously alive and dead at the end of that minute. Only by opening the box would the state of the cat's health be ascertained.

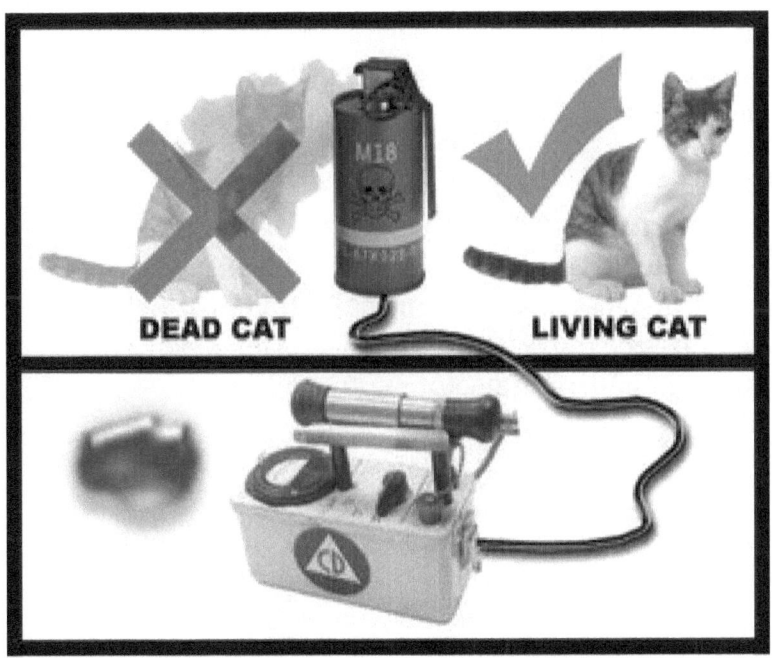

Figure 7: Schrödinger's cat may be simultaneously alive and dead until we look in the box. (image taken from Wikipedia).

The actual experiment with the cat would almost certainly not be possible in practice, as quantum effects such as state superposition in general are not thought to be noticeable on the large or macroscopic scale (exceptions to this rule include lasers and superfluids). However, the equivalent has been done for positive ions (atoms missing some electrons). Positive ions are also known as Cations. So the humorous title of an article on this experiment was 'Schrödinger's Cation' (Horgan 1996). Although nothing compared to the dead-alive animal, this experiment was quite remarkable in that this effect had not yet been seen for particles so large. Atoms, after all, may now be photographed using the latest scanning tunneling microscopes.

Note that I am not going to argue strongly for quantum effects explaining everything about consciousness. But later I will argue against the opposite extreme, namely trying to understand the brain/mind as if the quantum revolution in understanding of matter had never occurred.

Another actual experiment which gives a glimpse of the weirdness underlying everyday reality is the Quantum Zeno effect. This effect reminds one of the phrase 'a watched kettle never boils'. This is quite literally true in quantum systems. Quoting from the Wikipedia article on the Zeno effect:

"One such experiment was performed in October 1989 by Wayne Itano, Dan Heinzen, John Bollinger and David Wineland at NIST (http://tf.nist.gov/general/pdf/858.pdf). Approximately 5000 $^9Be^+$ ions were stored in a cylindrical Penning trap and laser cooled to below 250 mK. A resonant RF (radio frequency) pulse was applied which, if applied alone, would cause the entire ground state population to migrate into an excited state... The ion trap was then regularly "measured" by applying a sequence of ultraviolet pulses, during the RF pulse. As expected, the ultraviolet pulses suppressed the evolution of the system into the excited state."

A variation on this theme is a seemingly paradoxical experiment in which a quantum computer can get an answer to a problem without ever doing the calculations (http://www.nature.com/news/2006/060220/full/060220-10.html). It is these effects which would make a quantum computer extremely powerful. Such practical applications are forcing scientists once more to face up to the basic weirdness of a universe which is fundamentally insubstantial in its construction.

One experiment which anyone can do to demonstrate quantum weirdness is one described by John Gribbin in his book "Schrödinger's Kittens and the Search for Reality". All you need are three bits of polarized glass, such as the lenses used in polarized sun-glasses. Now the classical explanation of how such sun-glasses work is that they cut out light waves vibrating in one direction, thus reducing glare off reflected surfaces. In the diagram below, the classical view is illustrated by the old rope and a fence trick:

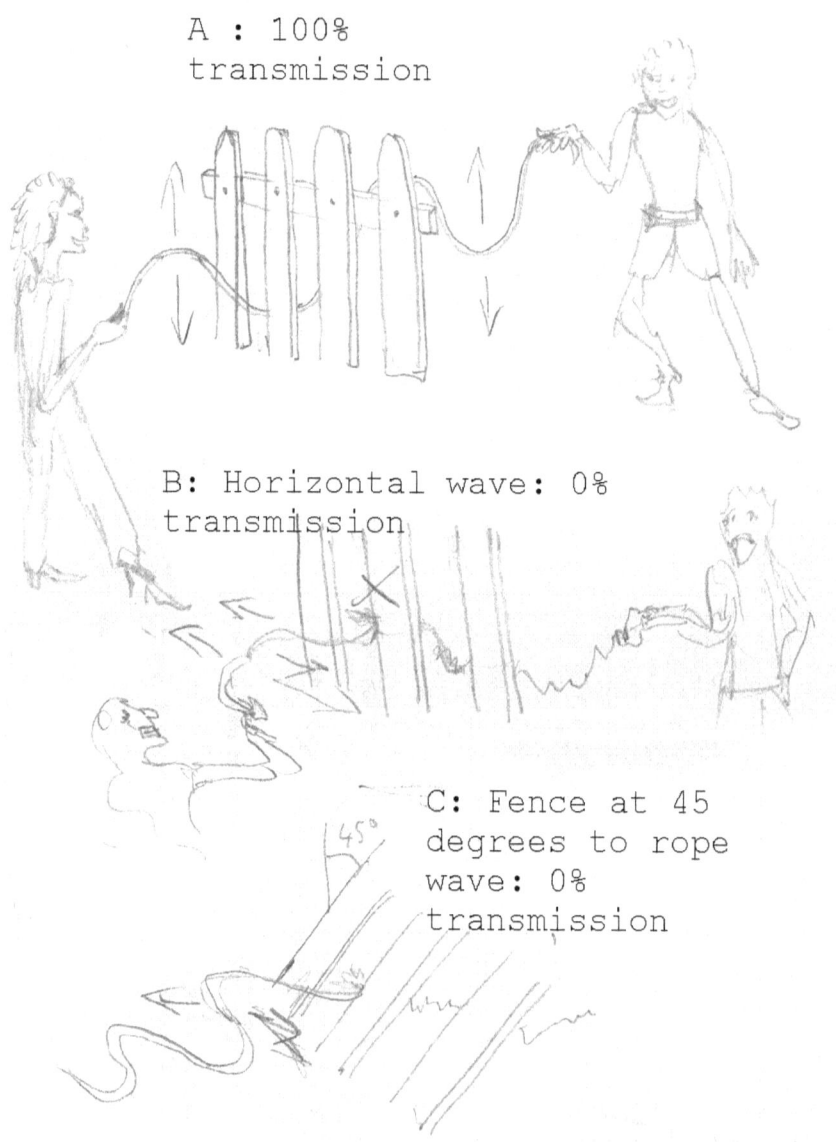

Figure 8: The classical picture of polarization of waves (image by author).

In the following picture, however, the quantum situation is illustrated, with Polaroid glasses and light waves. Polaroid glasses are very fine grids – like very thin barred fences: they have to be thin to block light vibrating in a plane perpendicular to the grid:

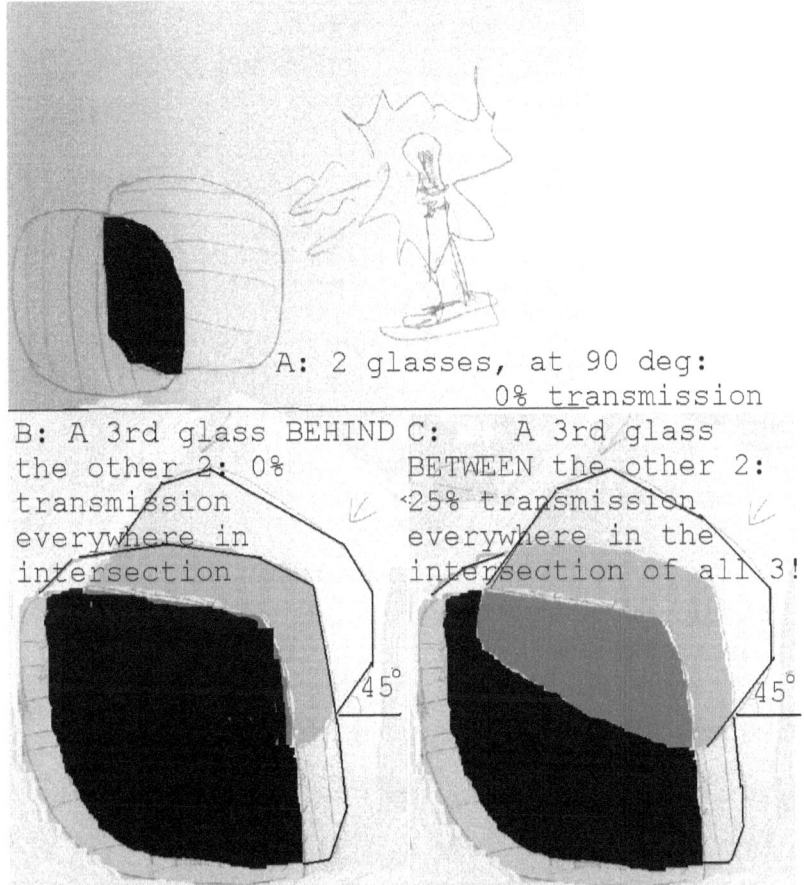

Figure 9: Polaroid glasses and quantum polarization (image by author).

Note that the situation in fig. 9 (A) is similar to the classic picture as shown in fig. 8 (B). I.e. for two Polaroid glasses perpendicular to each other, no light should get through, as it blocks light waves vibrating in the horizontal and perpendicular directions and all directions in between. Also, fig. 9 (B) corresponds to the classical situation in fig. 8 (C) and (B) at the same time, i.e. waves going at 45 deg to the horizontal will be stopped by grids in the horizontal and vertical directions. Note that in fig. 9 (B) the 3rd glass is slipped behind the other 2.

However, fig. 9 (C) is a shock to the classical picture. Simply by inserting the 45 deg glass BETWEEN the other two, we suddenly get some light trickling through – less than 25% of full intensity. Such a picture is unheard of classically – how can simply putting the 3rd glass between rather than behind the other two make such a difference? The answer is that in quantum mechanics, i.e. reality, we don't have real waves at any given time – instead we have ghostly probabilities of the photons vibrating, say preferably in the vertical direction. But this does not mean that we will always measure the photon to be vibrating in the vertical: it may have a 70% probability of doing so. That means that 7 out of 10 times we would usually find it really oscillating in the vertical plane, if we were to measure it. However, 3 times out of 10 we could find it vibrating in any other plane – with lower probability the further we deviate from the vertical. Thus in fig. 9 (C) if the first glass is with vertical grid lines, this means that photons come through with 70% probabilities of oscillating in the vertical plane. They then encounter the 45 deg grid and the quantum measurement process says we collapse the wave function or probability onto the 45 deg grid. There is indeed some chance of them being found to oscillate in the 45 deg plane. Thus some of the wave functions collapse onto the 45 deg direction. They are now reset with 70% probability of oscillating in the 45 deg plane so that when they encounter the horizontal grid, there is still a residual probability of them oscillating horizontally. That is then the light that gets through. We can also see that if we placed the slanted grid in front of the vertical and horizontal ones, we do no better than collapsing the wave functions of the 'vertical' (with 70 %) photons onto horizontal. But the probability as far away from vertical as the horizontal, i.e. 90 deg away, is essentially zero, so no light gets through.

So this remarkable experiment, which anyone can do at home, is a concrete proof that light and matter are not made of fixed classical objects, but of ghostly wave functions and their probabilities. And I have personally done this experiment at home many times, with the greatest of pleasure: in fact instead of breaking pairs of sunglasses, I ordered un-cut squares of Polaroid glass from my local opticians: these are the squares out of which they cut the sun-glasses. With these squares it is easier to line them up close, to maximize the intersection – which gives a more satisfying effect. Mmmm... wonderful!

The Second Revolution: Einstein's Thoughts on Relativity

Not long after Planck's bombshell, Albert Einstein revealed his Special Theory of Relativity. Like Planck's Quantum Theory, he felt himself forced to introduce a radical departure from classical Newtonian physics because of some experimental data on light that could not be otherwise explained. In this case, the observation was that if one measured the speed of light in a vacuum, one always got the same speed, i.e. about 299,792,458 meters per second. It didn't matter at which speed one was traveling relative to the source of light. In Newtonian physics, if you were traveling at 792.458 meters per second, light traveling in the same direction should appear to have a speed of 299,000,000 meters per second, obtained by subtracting one's own velocity from that of light. However, instead we observe no change in the measured velocity of light, c. This c is thus considered to be one of the fundamental constants of physics.

In order to always get the 'right answer' for c, it was necessary for space and time to bend somewhat, according to Einstein. If one was traveling in a spaceship going close to the speed of light, and looked back at Earth through a powerful telescope, one should see the people there moving extremely slowly. They, too, seemingly paradoxically, would see you moving around very slowly in your spaceship, and clocks there would appear to go slow relative to those on Earth. They would also notice the spaceship and you becoming distorted. You would appear to shrink in the direction of motion. Again, these effects seemed to depend on the observer. Everything was relative in relativity. It was relative to the observer.

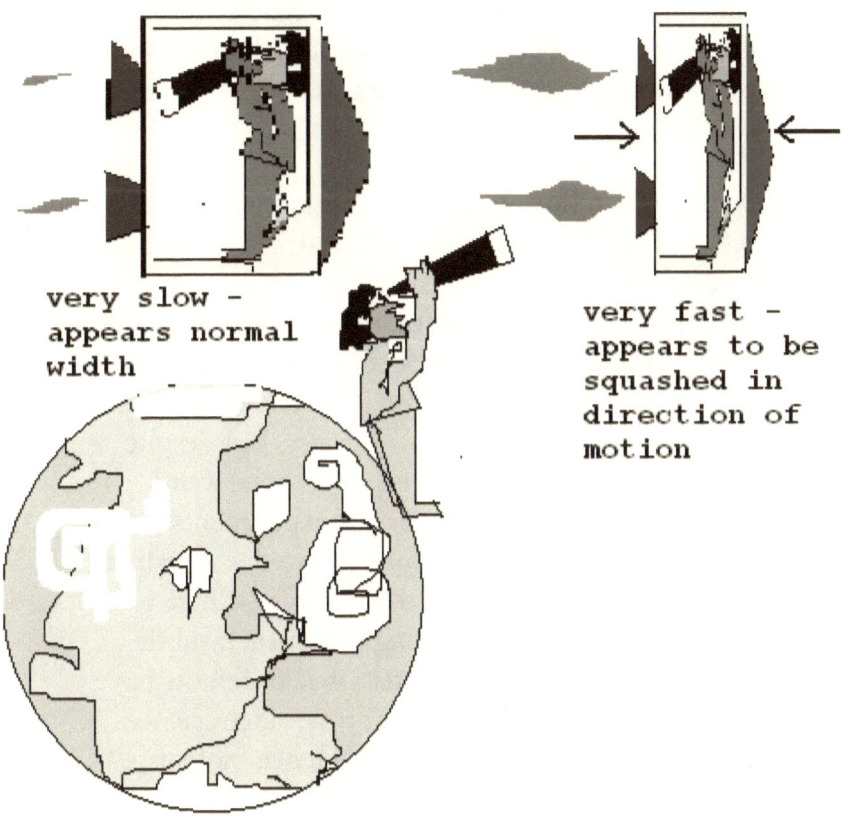

Figure 10: The spaceship appears to be 'squashed up' in the direction of motion as its speed approaches the speed of light. (Drawing by author).

Another effect on an observer in relativity would be if she were to travel near to the event horizon of a black hole. This effect involves Einstein's second major theory, the General Theory of Relativity. The closer she came to the horizon, the slower her watch would appear to her colleagues in the 'mother ship'. Eventually the watch would appear to freeze and she would slowly fade away. To her, though, things would appear rather quicker and more tragic.

So we see that in both of the revolutionary new branches of physics, quantum and relativity theory, the observer came to play a role. This almost heretical break with the Newtonian tradition of concentrating on the 'object' or third person aspect of things was one of the most surprising and for many old school physicists unacceptable aspects of the new physics. But despite such a genius as Einstein arguing against the weird quantum effects, nearly 100 years of experiments have just confirmed quantum weirdness, as well as Einstein's own relativistic distortion. So the observer is here to stay in physics. The curious thing is that such subtle revolutions in thought in one area of science now have trouble filtering through to other areas, since science has become so compartmentalized. Hence biologists, insofar as they consider physics at all, continue to work almost entirely within the realm of classical physics. They do not normally see the necessity of considering quantum effects (for example in neuroscience, quantum processes are usually neglected, as they are in evolution, where they could, for example, help to explain how the first RNA or DNA was formed) or the relativistic one. Yet the atoms that make up the material aspect of things have this ghostly quantum existence and on the large scale relativity is all-important. So it is only reasonable to expect some quantum effects even on the large scale. Lasers are a famous example of this, as the laser effect is a purely quantum one – classical physics cannot explain it. We will return to this subject later in the discussion of consciousness. Let us first look at attempts to reach a 'Theory of Everything' which unites relativity and quantum mechanics.

"Theories of Everything"

So far we have two successful theories for describing the world. General relativity describes the world of the very large, i.e. galaxies flying apart or spaceships returning from the moon or Mars, while quantum mechanics describes the world of the very small, i.e. protons, electrons, photons etc. at the sub-atomic scale. Both theories work wonderfully at their respective scales. The problems arise when we seek to unite the two to have one consistent 'Theory of Everything' (or TOE), since it turns out that quantum mechanics and general relativity are not so easy to unite.

The best known of the prospective TOEs is String Theory, known sometimes as Superstring Theory or M-theory. The basic idea is that sub-atomic particles are described as consisting of much smaller one-dimensional objects known as 'strings' (though unlike normal string they are infinitely thin). When this theory first became popular in the 1980s, there were great hopes of it providing an elegant solution to the problem of uniting GR & QM. However, complications soon arose as it became apparent that there were many possible solutions. Now, more than 20 years later, the problems have multiplied but many physicists continue to work on it as they have staked their reputations on it being 'the only theory'. In a series of books and articles recently, the problems have been described in some detail. For example, in his book "The Trouble with Physics", Lee Smolin (Smolin 2006) made a fairly comprehensive listing of the problem areas and Woit in "Not Even Wrong" suggested that String Theory was a myth.

First, String Theory has as yet to make a prediction that has been experimentally verified. This is an important point as every physics theory in the past stood or fell on its ability to make concrete predictions that could be tested by experiment. There are admittedly some predictions, but they need such high energy experiments that it is either beyond the resources of planet Earth at present or inherently impossible to perform them. This is a symptom of how String Theory drifted ever further into fields of pure mathematics, losing more and more the connection to reality.

Second, String Theory is basically a background dependent theory. This means that it assumes that space exists as a sort of blank background upon which the particles appear as actors on a stage. There is no obvious way to fix this at present, as it is inherent in the way the basic strings oscillate – they do so against a background of space. But the problem is that in Relativity, space-time is part of the action, bending when gravity is strong etc. and so it is not just a stage for the actor particles to play on.

Third, 'Supersymmetry' is needed, This is a property of particles that says every particle has a corresponding 'supersymmetric' partner. If the particle is a fermion (e.g. electron, Proton, Neutron, with odd multiple of spin 1/2), its partner is a boson (even number of spin ½, e.g. photon, weak bosons W, Z) and vice-versa. Thus for the electron there is the 'selectron'. These s-particles are very heavy, but at least some of them should have been seen in accelerator experiments by now. If they are not seen in the LHC (Large Hadron Collider), then String Theory will be in even deeper trouble than at present.

There are several other objections to the theory. Smolin, in his book, also describes how there are a huge number of possible String theories – i.e. uniqueness is not achieved. Also, for a viable theory, some extraordinary mathematical contortions have to be assumed, which look anything but natural. All these problems make String theory look rather ugly and not the elegant and simple solution it once promised to be.

However, there are other theories beside String that require supersymmetry, so even if supersymmetry is detected by the (soon to be commissioned) Large Hadron Collider it would not necessarily be a vindication of String theory. The main rival to String Theory as TOE is Loop Quantum Gravity (LQG), as developed by Smolin, Rovelli and others. There are in fact other Quantum Gravity theories that resemble LQG. They all have in common, though, that they are background independent: i.e. space is not a neutral background but part of the particle-force setup. In LQG, space-time is quantized, i.e. there are quanta or 'atoms' of space arranged in a sort of lattice, instead of the smooth, featureless space of String theory.

Note that LQG is finding it hard to attract physicists to work on it, as String Theory has a near monopoly on funding in this area. And though Strings have faced several crises that shook the field to its core, physicists tend to stick with it as the initial attraction still holds, despite all the ugliness of the fixes: I.e. the Standard Model emerges from the theory in a natural way, as does a particle with spin 2, i.e. the graviton. This early success caught the imagination of the physics community and soon almost every theoretical physicist was working on it. But after 20 years of intense research, the early promise has failed to crystallize into predictions of experimental results. Thus verification is far away, and there are other major problems such as lack of a unique theory. It is true that some wonderful new mathematics has emerged from the search for a String TOE, but just because the initial theory was 'beautiful' and elegant in a mathematical sense does not guarantee it being true. This was the case for Ptolemy's theory of epicycles to explain the motion of the planets: planets sometimes stop, reverse in their path across the sky and then continue their original direction. Ptolemy could model this by a complex system of interlocking cosmic clockwork. Unfortunately, though it was beautiful, it was also utterly wrong. Kepler, Galilieo, Newton & Copernicus put forward a new theory that was in some ways more ugly (the planets orbited in ellipses, which lacked the classical beauty of circles), and yet it was true.

A dark horse, rapidly brightening, in the area of TOEs is the theory of the German physicist Burkhard Heim (1925 – 2001) and associates, which recently shot into the limelight again with cover stories in New Scientist and an exciting connection to a possible revolution in fundamental physics. Like LQG, it is a quantum gravity theory.

Through a chance meeting with Heim's associate Illobrand Von Ludwiger in 2000, I became involved in the revival of this old theory which holds great promise of becoming the major rival to String Theory for producing a 'Theory of Everything'. I learned of Heim theory at a series of presentations in German by Von Ludwiger at a time when information on him in English was scarce. I was impressed and so started articles on his life and theory in the English version of Wikipedia. I must say that I was surprised at the effort that was necessary to defend these entries against the threat of deletion by Wikipedia editors either too lazy to investigate the theory for themselves or convinced that String Theory was the only TOE candidate of note. It was then that I appreciated the descriptions by Smolin and others of the difficulty in breaking the monopoly of String theory as a TOE.

But in 2005 Heim's main collaborator Walter Dröscher, together with Jochem Häuser, won a prize at a major meeting of the principle international association for Aerospace, AIAA, for their presentation of how an extended version of Heim theory could lead to a new space propulsion system. January 2006 New Scientist ran a cover story on the theory, emphasizing how it might allow faster-than-light travel, after which the threat to suppress the Wikipedia article eased. There is nothing like publicity in the face of censorship! A good deal of excited discussion about the theory took place on some internet forums, notably the thread http://forum.physorg.com/index.php?showtopic=4385&st=0.

Heim's theory says that the quarks are really condensations in a 6-dimensional process or flux responsible for giving particles their mass. In contrast to String Theory, Heim does make several testable predictions. Firstly, it actually predicts the masses of the sub-atomic particles to an accuracy not yet achieved in the 'Standard Model' or in String Theory. If the mass formula can be confirmed by theoretical physicists, the implications would be tremendous. At the moment, however, only a few scientists are studying the theory: it needs some of the top minds in physics to unravel the complicated multi-dimensional arguments of the eccentric Heim, who worked in isolation due to major physical handicaps from injuries received in an explosion in World War II.

The second major prediction of Heim is for 2 additional forms of gravity, one of which is quintessence or dark energy, and the other involves a set of 3 particles, dubbed 'gravito-photons', one of which involves repulsive or anti-gravity. The theory also shows how photons can be converted into gravito-photons by a spinning magnet set-up, which would, if properly constructed, produce an anti-gravity effect.

Many were wondering about the prediction of Heim that an engine whose core was a spinning magnetic setup could produce anti-gravity, when a few months after the New Scientist story appeared, the Austrian physicist Martin Tajmar and co-workers announced via the ESA website that they had produced the first real indication of actual artificial gravity of the type Heim had predicted. Their study had been funded by ESA, the European Space Agency, who recognized Tajmar and his lab at Seibersdorf in Austria as a thorough and conscientious researcher. Dröscher and Hauser showed that Heim's equations had predicted Tajmar's results, and what's more, they showed how by re-arranging Tajmar's setup, a truly anti-gravity effect could be achieved, i.e. parallel to the axis of the spinning ring or disk (Tajmar had produced a force tangential to the spinning magnet). So if this extension of the experiment really were successful as predicted, it would be further evidence that Heim's theory was on the right track. Needless to say, such anti-gravity would be a major new technology of the 21st century. In July 2007 a group in Canterbury, New Zealand, produced the first tentative confirmation of Tajmar's results, though they were too cautious to admit it themselves. The satellite Gravity Probe B, whose mission was to confirm certain predictions of general relativity, may also have produced corroboration of Tajmar's results, since an anomaly in the measurements of one of its instruments, which involves spinning superconducting metal balls, may best be explained by the gravito-magnetic effect of Tajmar. Other groups are also working in this area now, and so soon it may be confirmed again and the technology may then start to 'get off the ground' literally.

In the table below, courtesy of I. Von Ludwiger, a comparison of the main features of String, LQG and Heim theories is given. If you find it too technical, just skip over it – the upshot of it is that Heim Theory compares well with the other two and so looks very much like a serious candidate TOE.

Table: Comparisons between LQG, String and Heim theories

Properties and features	String theory	Quantum geometry (LQG)	Heim theory
Description area	space-time R4	R4	R6
Fundamental objects	Space, time, strings & branes	Spin network, spin-foam	Surface quanta (metrons)
Number of spatial dimensions	9 or 10	3 (more possible)	3
Number of imaginary dimensions	1	1	3 (or 5)
space-time as a	yes	no	No

background metric			
Modification of general relativity	yes	yes	Yes
Modification of quantum theory	yes	yes	No
Conceptual Fusion of quantum theory and general relativity	no	yes	no (or possible)
new physical principles needed	yes	no	No
Nature of matter	Excitation states of strings/branes	States of the spin networks	Cyclic exchange processes of maxima of structure compressions
Explanation of standard model of matter	only tentative	not required, not possible	with extension to 8 Dimensions
Unknown elementary particles predicted	yes	no	neutral electron
Explanation for dark matter in the	maybe	no	Probably

53

universe infinity problems in the formalism	no (?)	no (?)	No
unification of quantization	not of space and time	yes	(under development)
Unification of the forces of nature	yes	no, but allowed	Electromagnetism and gravitation
Super-symmetry required and presupposed	yes	no	No
Uniqueness	no (many string vacua)	no (ambiguity in Hamilton operator)	Yes
Existence of many other universes	possible	uncertain	Yes
Explanation possible for beginning of universe	Big Bang model	Big Bang model	Zero Matter origin model
Explanation for entropy of black holes	limited	yes	not yet examined
Explanation	possible	possible	not necessary

54

possible for cosmic Inflation			
Description of scattering experiments	yes	not yet achieved	not yet calculated
Explanation of fundamental physical constants	no	no	Yes
Contact with low energy physics (everyday world)	only tentative	only tentative	Fully
Verifiable predictions	only tentative, partly falsified	only tentative	Masses and lifetimes of elementary particles and resonances & anti-gravity force via 'gravito-photons'

Extended table (von Ludwiger, 2006, private communication) after R. Vaas: The duel: Strings versus loops, in "Bild der Wissenschaft", 4, 2004

Some progress is being made in verifying the theory via Tajmar and others' work with spinning magnet systems. The mass formula is also being tested – notably in the thread on physorg.com listed above, where several physicists and computer experts have produce different implementations of the mass formula. One of these physicists, John Reed, has done more than others to relate their implementations of the formula to Heim's books and other writing. At the time of writing, I am working as a sort of liaison between the Heim Theory group in Europe and Reed in the USA. The exchange has already shown up several minor typos in some of the equations of the mass formula. Also, Reed stopped work on the mass formula for a while as he found that early versions had used a mysterious 'A' matrix, containing data from particle physics experiments. This was reported in the Wikipedia article and elsewhere and seen as a triumph for the anti-Heim lobby. However, the HT group could eventually show Reed that the later version of the formula, from 1989, was free of the A matrix. Reed agreed to this, retracted his objections and started work on the formula again. This work is ongoing and rather exciting.

Summary

We have seen that the two great movements in new physics of the 20th century led to an inclusion of the subjective observer in what had previously been a purely objective description of the world.

The growing importance of the subject also has implications for our view of human nature. In a world where magazine articles and popular science books constantly seek to remind us that free will, the mind or the self are mere illusions, any voice in the wilderness protesting against the apparently crushing weight of neurological evidence needs all the ammunition it can get. Thus the subjective aspect of the new physics is a useful argument against the nihilistic tone of those anti-human articles which toe the party line of the materialistic 'consensus' that has become the new status quo.

This dominant materialism and accompanying tendency to eliminate the subject from science has also had catastrophic effects in medicine. As long as a patient is just seen as a bio-machine in need of adjustment, the negative effect on the old doctor – patient relationship is enormous. This is part of the reason that doctors simply prescribe endless pills for their patients. But even in medicine the tide has begun to turn. The power of the subject to interact with the 'object' of the body is being at long last recognized as a powerful force for healing. What was once patronizingly dismissed as the mere 'placebo' effect is now acknowledged to be a manifestation of mind-body interaction. Countless studies have now shown that the immune system is improved by positive thinking, music or laughter. Traditional medicine such as Chinese or Tibetan systems or shamanistic rituals has been shown to have an effect on at least some illnesses. The holistic nature of the traditional approach to medicine is something that has developed over thousands of years and should not be thrown away simply because modern western medicine has made breakthroughs in a number of areas. Sometimes the claims for the traditional methods are exaggerated. The best solution is probably a combination of modern and traditional techniques.

Yet another ill effect of the erosion of the idea of free will and the self is in the area of the law. The obvious defense in a trial could be that the defendant has no free will or selfhood, and thus is not responsible for her actions. This sort of thinking, if it is established as a precedent, could cause immense damage to the structure of society.

It is sometimes amusing to see the rather lame efforts of certain journalists and neuroscientists to ascribe all religious feeling to certain areas in the brain. I will also have more to say on that later. As remarked at a recent philosophical meeting that I attended, such claims make for good copy around Easter, every year. The next Easter the articles appeared again, prompt as ever. But as usual these articles are rather shallow. Theologians have long ago begun their counter attack and their arguments are a bit deeper than those of the magazine articles and books by Dawkins & co.. But of course that doesn't make for as good copy.

So, we have seen that the subject has begun a triumphant come-back in the 'hard' sciences and is slowly working its way back into the 'softer' sciences of medicine and biology. Next we shall see that the attempt to duplicate the mind has not been as successful as its proponents had hoped half a century ago.

The failure of Artificial Intelligence.

In the last chapter we saw that the new physics has re-instated the subject in the realm of fundamental science. Developments in neuroscience and Artificial Intelligence in the last 50 years have similarly implied a failure to fulfill the promise of the computer model of the human brain/mind. The latter model had grown out of the mechanistic view of reality whose modern form started with Rene Descartes. The latter had divided the world into a Res Cognitans (thought stuff), i.e. the human mind ('I think, therefore I am') and everything else (Res Extensa). Thus in this view animals and plants were nothing but mechanistic machines. Only man was an exception, with his reasoning intellect. Later, even man's intellect or mind would be swallowed up by the machine theory. As stated by Penrose and John Searle, each age compares the mind to the most complex technological machine yet constructed. Thus in the 18th and 19th centuries, the comparison was with clockwork. Many were fascinated by the clockwork dolls, which could be programmed to perform complex tasks. In the same way, the 20th century came to compare the mind with a computer, as this was the most advanced machine of the age. Modern psychology has indeed shown some similarities between the way the mind-brain works and the way a computer works. E.g. in experiments where people were to identify different types of birds or animals, it took them marginally longer to identify rare birds, like the kiwi or platypus. This showed that the brain does indeed store information in a hierarchical representation. Thus cognitive abilities do indeed resemble those of computers. But the resemblance only goes so far, just as the clockwork resemblance only goes as far as muscles resembling springs etc.

The comparison of the mind/brain to a computer developed into the scientific study of Artificial Intelligence (AI). Major researchers in AI have always predicted that computers would soon become conscious. Thus the date for this great event was originally set for 1970 and was later pushed back to 1990 and then 2000, but here we are years after that with no sign of computers with the same cognitive capacities as the human subject. Now the prognosis is for a conscious machine in 20 years – yet again.

 Roger Penrose's book 'The Emperor's new Mind' made the comparison of AI with the fairy tale of the naked emperor: no scientist dared to point out that the ruling AI theory was devoid of any strong proof. In particular, Penrose points to the fact that although computers are very good at certain repetitive tasks, they are far from being able to replace or reproduce the activity of human brain/mind subjects. And this is not merely a question of computing power, as suggested by many commentators, e.g. Malik (2001). It is more like comparing chalk with cheese.

Chess and computers

A good example of this was the artificial chess problem posed to the supercomputer Deep Blue, cited in Penrose's 'Shadows of the Mind' (2000). To any human novice in chess, it was obvious at a glance how white could achieve a draw. Deep Blue, however, after much deep CPU thought, took the obvious bait of taking the rook with her pawn (on the left of the board) and lost (see Figure 9). A human player sees at a glance that by taking the inviting rook, the impasse of one wall of pawns facing another is broken, and black can then come through the gap with her other players, defeating white. Thus all white has to do is move her king back and forth — black's powerful pieces are all incapable of taking a white pawn if white doesn't take the rook and so are unable to utilize their potential. If you play chess at all you see that immediately. Deep Blue, however, remains in darkness until a programmer comes along and puts in a lot of additional code to deal with this special situation. But we didn't need special code, as our general understanding of the *meaning* of the game means that we are flexible when another combination presents its meaning to us.

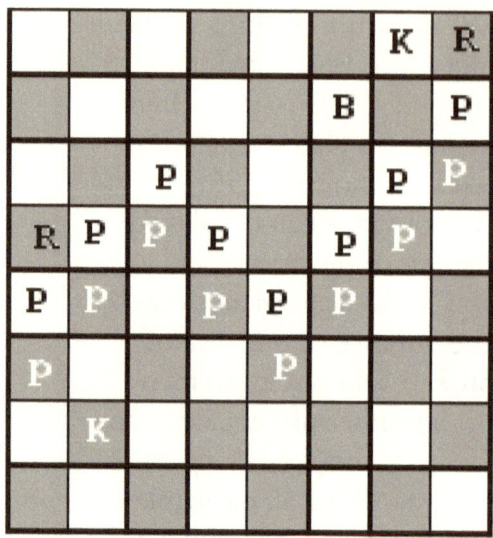

Figure 11: Problem given by William Hartston, from a paper by Seymoure and Norwood (1993) and cited by Penrose (2000). It is White's turn to move, and her task is to play for a draw. Easy? Not for a supercomputer!

Thus Deep Blue's victory over Kasparov (see Figure 10 below) was due to a different form of brilliance to that of humans - namely brute algorithmic computing power, as opposed to the non algorithmic insights of the human brain/mind.

Figure 12: Kasparov playing Deep Blue in 1996 (picture taken from Wikipedia).

That is, computers simply don't understand the meaning of what they see, which is a feature of intuitive subjective intelligence. A similar point was made by Searle (1980) with his Chinese Room analogy, in which a westerner, knowing no Chinese, blindly followed a set of rules to always give the right answer in Chinese to questions posed in that language. But someone who really has learnt Chinese also *understands* what she is doing whilst answering the questions. This thought experiment has been attacked by many writers and defended by others. One of the critics was Steven Pinker who (Pinker 1999, P. 95) stated that his view was that "Searle is merely exploring facts about the English word *understand*." However, my own take on the Chinese Room is that it could be made clearer by something which Searle may have omitted. Namely, what is missing is the *qualia* or inside view of a consciousness that understands the meaning of the questions. One would have to imagine a ghostly light floating in the library – this would be the being that understood the questions and gave the appropriate answers. The fact that there is no such light leads to Searle's puzzle and the reduction ad absurdum of his critics in attacking it. One reason that Searle never really emphasized this aspect of the problem is that he himself considers the brain - consciousness link to be analogous to the stomach – digestion link. But of course this misses the point, as digestion is just another objective process, while subjective consciousness is the only non-objective process we know: i.e. the only view from the inside outward instead of from the outside inward. In this sense consciousness is the light shining in the library of the brain. See the section on subjective consciousness below for a fuller discussion of this theme.

'Go' and computers

Another game that is of comparable complexity to Chess is the ancient oriental game of Go. For those unfamiliar with the game I reproduce a sample in figure 11 below. In Go, the pieces or 'stones' do not move around. At the beginning, black has 181 stones and white has 180. Together that makes up 361 = 19 x 19, as the board is a 19 by 19 grid. A turn consists of simply placing a stone at some position on the grid, where it will stay unless captured by being separately surrounded by the opponent or being attached along horizontal or vertical lines to a group which is totally surrounded. Black has one more stone at the beginning as she plays first. The rules are simple, but on a grid of this size complex situations arise very quickly.

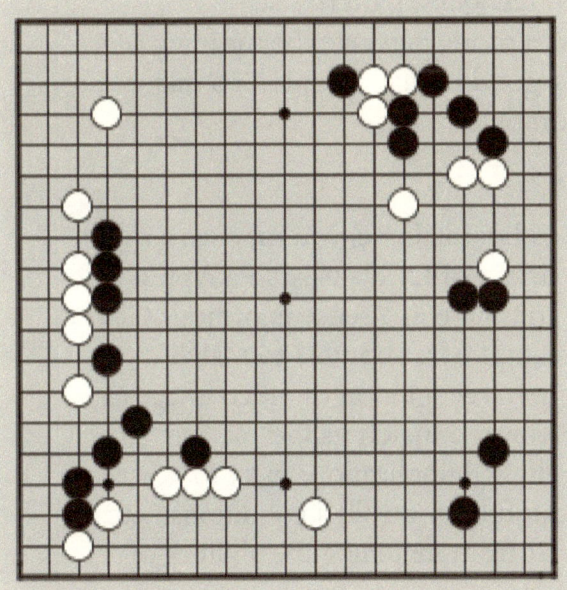

Figure 13: Sample Go board with game in progress (image from Wikipedia). Millions of human Go players can still beat the best programs on the most powerful computers!

The advantage of the human style of thinking as opposed to the most powerful modern computers and associated programs is more obvious in Go. Unlike in chess, where only a handful of masters can occasionally beat the best computers, no computer has beaten a master of Go yet. In fact, millions of Go players are still better than the best computer! This was seen as a challenge to AI, so that there has now been almost the same amount of effort invested in developing good Go software as there has been for Chess programs. But the human ability to recognize patterns, see connections and even more, perceive the meaning of a situation still gives a good human Go player a major advantage over a computer.

If you look at the Go board above, you may see a slight resemblance to the arrays of CCDs used in a typical Camcorder. These arrays are in a sense enormous Go boards. Thus the attempt of computers to match human ability in pattern recognition of video images is analogous to the attempt to match its Go playing ability. While a few limited applications exist to recognize faces in carefully controlled conditions, etc., computers still are far from matching the abilities of the human sense of vision.

Language and Computers

Let me note in passing that the philosophers and neurologists beloved of the media only seem to delight in describing deficits in the human brain. Thus we read of 'The Man Who Mistook His Wife for a Hat' (Oliver Sacks 1985) and of patients with other brain injuries, until you think the brain consists only of deficits and bugs. Then other writers praised in the media, such as Daniel Dennett, when describing vision, love to dwell on the 'faults' and inefficiencies of the process, such as the fact that we have sharp vision only in the centre of the field of vision. Dennett (1993) asks one to describe something on the periphery of the field of vision and then triumphantly deduces that we only have a very limited sense of peripheral vision. Yet we fool ourselves into thinking that we see the whole room, since we scan it and recall where things are, thereby thinking we see more than we do etc.

I personally prefer a cognitive scientist like Bernaard Baars' description of vision as something of awesome complexity. When it works properly, which the likes of Dennett are not too fond of dwelling upon, vision and intelligence as a whole are staggeringly complex. As Baars (1997) points out, there are 40 or more areas in the brain concerned with analysing the image projected on the retina. The process of recognising an object involves multiple layers of image processing of form, colour, movement, orientation etc. as well as links to memory and decision processes. In the example of reading a paragraph, we scan quickly many complex shapes and perform, all subconsciously, ultra-efficient analysis of the meaning of words, word groups and even sentence groups in a fraction of a second. Then in milliseconds we have summarised the main idea in the paragraph so that if someone asks you suddenly to say what the content was, you will be able to give a synopsis of the main ideas, without being able to recall the actual words used. In the process of doing this we have resolved several ambiguities in context and extracted meaning in a way that defeats the best computerised text analysis programs.

When one considers this complexity of reading or having a conversation, one understands why it is forbidden in most countries to telephone with a mobile whilst driving, since with so much computing power directed to the conversation, sometimes too little resources are left to attend to the driving safely. That again shows that the subconscious, though good at running routine tasks in a routine way, finds it difficult to deal with the unexpected, such as a bicycle suddenly swerving in front of the car.

As Baars and others point out, our conscious awareness is at the tip of a subconscious iceberg of immense complexity. Yet consciousness is like a customer taking delivery of user friendly goods, unaware of the mountain of subconscious programming underlying the words she is reading or the painting she is admiring. Consciousness always floats on this sea of complexity and is nevertheless in control. In learning a new task, the conscious mind is in full control – e.g. learning a new computer game. But once learned, the ability is shunted onto the subconscious and may in future be executed automatically with only minimal conscious intervention. This is what gives the human mind such power: As Baars says, we can switch between utterly different tasks – driving a car, writing a story, doing maths, getting a bottle of milk from the fridge, singing, dancing etc. effortlessly – to learn something new all that is needed is to direct the conscious mind toward it until it becomes automatic. No conceivable program has this flexibility, not now and maybe never. So this is the attitude I will most frequently be taking in this book – not of knocking the brain-mind and poking holes in its abilities, but of praise and awe and wonder at the amazing abilities a healthy brain – i.e. that of the vast majority of people who are fortunate not to suffer from brain lesions - possesses.

Another point about Baars compared to more negative analysts such as Dennett is that although he also describes the results of brain injuries or lesions at great length, he also points out the incredible flexibility of the brain/mind in adapting to the restriction caused by the injury. Where a robot computer would crash utterly after such an insult to the system, the brain/mind (note I always say brain/mind and not simply brain as it is not certain that the equation 'brain = mind' holds good, as we shall see in later chapters) often finds a way to work around the injury, and often in time will heal it by having other brain areas take over the missing function. This is an example of the plasticity of the brain. One of the first well documented cases of such plasticity was that of the railroad worker Phineas Gage, who in an explosives accident in 1855 suffered a spike piercing his brain from his eye to the back of the head. It greatly damaged his brain's frontal lobe and though his personality was altered in that he became more irritable and aggressive, he was able to live an independent life.

Thus, though great strides have been made in computer translators, it has proved difficult to go beyond a certain level of expertise, as human speech has turned out upon deeper analysis to be deeply rooted in human culture with all its complexities and myriad links and dependencies. Without understanding the *meaning* of the actual sentences, it is sometimes impossible to decide which of several senses to employ for a word like bat – e.g. we know that the bat in 'the bat was in the cave' is almost certainly different from the one in 'the bat hit the ball, sending it down the field'. We know from the sense or meaning of the context which 'bat' is implied.

The Challenges to Artificial Intelligence

So there seems to be something special about the brain-mind system compared to a computer. This argues against the idea that our minds are just aspects of a lump of matter. There is something anonymous and pejorative about the idea of being any old lump of matter. Anyway the idea of a simple lump of matter isn't what it used to be. First, 'matter' is not the hard and fast stuff it used to be, as we saw in the previous chapter. Break up the lump of matter and you find ghostly probability patterns - the stuff of which dreams are made. Secondly, saying, as many ultra-materialist philosophers do today, that consciousness is nothing but the 'totality of feelings corresponding to being a lump of matter' reminds me of Colin McGinn's book on mind and matter. He (Mc Ginn (1999)) ridicules rather skilfully the idea that we are nothing but a 'computer made of meat'. Particularly brilliant I found his description of space and mind - every phrase and idea relates to space, but thought is not of space - consciousness is not located in any one place and does not have spatial extent. We will return to this in the chapter on subjective consciousness. Suffice it for now to point out that it is this mysterious aspect of consciousness that poses the greatest challenge to AI.

Another challenge to AI arose in the 1990s when Roger Penrose brought out his bestseller "The Emperor's New Mind" (Penrose (1990)). In this book, Penrose pointed out how the promise of AI had not been met. Yet computer scientists, psychologists and cognitive scientists frequently spoke as if the development of a truly intelligent computer or robot was just around the corner. Yet every time a deadline approached for the materialisation of this wonder machine, the deadline was pushed back another decade or two. Penrose claimed that the main reason for this was that machines 'think' in digital algorithms, while human thought often, if not always, proceeds differently. As we saw in the examples of Chess and Go, the way we play those games appears to differ radically from that of computer programs.

Penrose used a proof that showed that no matter how many digital algorithms one can imagine, there is always one that falls outside one's system. This proof, analogous to that used to prove Gödel's famous theorem of the incompleteness of any set of mathematical theorems based on a given system of axioms, was criticised by others and defended by Penrose in the sequel to his first book, 'Shadows of the Mind.' Whether or not one agrees that he was right with this in showing how profoundly limited computer programs are, the basic idea is very strong. The very fact that computers have reached a complexity exceeding the brains of some animals without showing the same abilities is already indicative of some additional ingredient needed for sentience. But another argument of Penrose's is also interesting; namely, that mathematicians as well as artists or musicians have moments of inspiration, where they grasp a mathematical truth without formal proof. The proof comes later. Penrose implied that at such moments it's as if the mathematician is in touch with the world of abstract Platonic truths. It is indeed remarkable how simple it seems to us to grasp meaning out of different contexts. This again is something like the process of mathematical or artistic insight, but on a much more humble scale. Yet just because our grasp of meaning is such an everyday tool it does not detract from its mysterious nature. It is in turn maybe related to the idea of sensory 'qualia' or the quality of experience in seeing a red rose or a sunset or hearing Jimmy Hendrix. We will return to a discussion of 'qualia' in the chapter on subjective consciousness. As it is such an unfamiliar concept to most people who are not philosophers, and yet so basic when one realises its nature, it will need a bit of lateral thinking to see the world from the 'qualia' point of view.

Before going on to the chapter on subjective consciousness, however, let me just point to another aspect of the AI debate. While it is true that some Japanese robots now are able to walk like humans in certain limited scenarios, we have to compare this performance to that of a baby gazelle born on the savannahs of Africa. Within minutes the baby must be able to run from that lion who has noticed the mother. So within seconds of being born it is already performing with amazing skill a twisting path to escape the lion. The walking robot, despite having programs of a considerable size and complexity, cannot move with such skill, integrating vision and motor responses in this marvellous fashion. This begs the question of how such complex instincts can be inherited with a genome that is rather limited in its information content. We will examine this point in the chapter on the genome.

Finally, even if in the proverbial 20 or 50 years we do create a conscious robot, well and good. However, to truly prove that it is conscious in the sense that we are it should show a grasp of meaning, play Go like a human, and be able to describe its reaction to a new Shakira song with such enthusiasm and conviction that we must believe that it can really 'feel the rhythm deep in its soul'.

The challenge to AI is therefore rather huge. First, computers must be made that have that flexible sense of thinking that allows consciousness to play over multiple alternative interpretations and instantly select the correct one. Next, they must be capable of moving around in the real world without stumbling over rocks, furniture and other obstacles. Next, they must have a master plan and a series of nested lower priority plans as we humans do. Most difficult of all for the robot, though, will be to convince us that it is really enjoying the sunset and is *subjectively* aware of colour, music, tastes, tactile impressions etc.

Subjective consciousness

As promised in previous chapters, we now turn to that aspect of consciousness that is least mechanical or quantitative. We must come to understand, in many cases by exercising a decisive leap in lateral thinking, what it is that is really special about the *quality* of consciousness or mind.

Let me first say something about how I personally came to realize what this 'subjective' aspect of consciousness is. Having trained as a physicist I was used to thinking of everything as a material system obeying the laws of physics. If you throw a ball to your dog or cat, it flies through the air in a parabolic arc due to the force of gravity. If it's a windy day the ball will be blown off course. That's another force acting on it – wind and air resistance. When a car drives off I know about the little explosions in the internal combustion engine and how they push pistons which transfer the motion to the wheels. If particles crash together in a particle accelerator I know why they spew out a shower of other types of particles. People trained in the 'hard' sciences like physics and chemistry are used to doing experiments and theory that treat everything as predictable systems of matter.

Penrose, Chalmers and the Brain/ Mind Debate

As a physicist, one tends to trust famous members of one's profession. Thus it took another physicist, Roger Penrose, to make me take notice of the fact that there might be another aspect to the mind than atoms moving around the brain. His blockbuster "The Emperor's New Mind" (Penrose (1990)), besides presenting his own arguments, served as a general introduction to a physicist or anyone else lacking in knowledge of some of the major philosophical debates. Thus more from following up the references than from the arguments in Penrose's book, I became aware of the on-going debate between the various schools of thought in the mind-brain debate.

The point which only gradually sunk in, and which was something like an illumination or revelation was the nature of subjective consciousness. David Chalmers is the philosopher who more than any other has popularized the explanation of this point. He likes to cite (Chalmers (1995)) the example of Mary, a scientist who has led her life in a sort of sensory deprivation chamber, except that the only sense of which she was deprived was that of color (ideally she should be from deepest Africa with utterly black skin). In other words, she lived in black and white. The point is that she is an expert on the theory of color – she knows all about the wavelength of light of each color of the spectrum, and which areas of the brain should process it etc. Only she has never experienced color herself. Then one day the door is opened and she can walk out into a colorful garden under a blue sky. She is in awe, because nothing of the theory she learned could express what she actually *experienced* when she saw her first red rose or green grass. In fact we now know that her brain would first have to form neural connections to learn to activate the colour sense mediated by various modules in the visual centre of the brain. But for the thought experiment it is assumed that such adaptation occurs instantly.

This is the essence then of subjective consciousness. It is the experience that defies description in words. I.e. it is ineffable. This is true of the redness of a rose or the sound of something by Mozart, Enya or the smell of a perfume. These are all sensory experiences, which we talk about and we understand something of the brain processes that gave rise to them. But we cannot capture the essence of the subjective experience in words. This was one of the revelations that had to sink in before I realized that the human subject (or dolphin or ape or parrot) was different from objective considerations of her as a member of society or an eco-system. There has always been a certain resistance by scientists of all sorts to accept this subjective aspect of things. Even Marx denied it and saw humans as defined by their social interaction rather than their individual experience. But the subjective aspect is now widely acknowledged to be an important side of experience – indeed these subjective sides of sensory experiences are known amongst philosophers and consciousness researchers in general as 'qualia'. To explain how the qualia and other elements of subjective experience arise from objective processes such as ions and molecules moving around the brain has become known, largely thanks to Chalmers, as the 'Hard Problem of Consciousness'.

The problem is somewhat complex. The Buddhist idea of the individual is that of the anatman or 'non-self': i.e. the self is illusory as this anatman is more a collection of processes than a concrete whole. This picture of the non-self may be seen as consistent with a tendency in Asian thinking for the individual, though recognized as a suffering, unique being deserving of compassion, to be subordinated to the whole. The western preoccupation with subjective consciousness is thus sometimes seen as a sort of sign of the individualism and conceit of westerners. On the other hand, the tendency of science, and in particular neuroscience, to deny the subjective dimension to humans and to treat such aspects as the self as illusions has been seen by many as an alarming development which undermines the self-respect of humans. The very notion of human nature is at stake, it seems.

Thinkers such as David Chalmers and Kenan Malik emphasize the fact that humans are subjects, as well as objects. Subjects are characterized by free will and sentience as well as the more fundamental aspects of perception such as the qualia.. These features of perception are the first indication that many aspects of mind are two sided, possessing a 'subjective' and 'objective' aspect. As Chalmers (1994) points out:

"When we think and perceive, there is a whir of information processing, but there is also a subjective aspect. As Nagel (1974) has put it, there is something it is like to be a conscious organism. This subjective aspect is experience. When we see, for example, we experience visual sensations: the felt quality of redness, the experience of dark and light, the quality of depth in a visual field. Other experiences go along with perception in different modalities: the sound of a clarinet, the smell of mothballs. Then there are bodily sensations, from pains to orgasms; mental images that are conjured up internally; the felt quality of emotion, and the experience of a stream of conscious thought. What unites all of these states is that there is something it is like to be in them. All of them are states of experience."

Philosophical Schools of Thought on Consciousness

The three main divisions in philosophy can be said to be materialism, dualism and idealism. The first of these maintains that all is matter and the mind, self and soul are all illusions. Idealism is the opposite – i.e. all is mind or soul, and it is matter that is the illusion. Finally there's the half-way house of dualism: there are both mind and matter, either interacting (mind acts on matter and matter acts back on mind), or non-interacting (i.e. coexisting but mutually exclusive).

Listening to the media, whether magazines, newspapers, TV or radio, one might conclude that the materialist thesis is dominant. Though what is in question here is maybe better referred to as physicalism, the idea that everything can be explained as a result of physical (from physics) processes. This term avoids some negative connotations associated with materialism. The reason that physicalism is the prevailing paradigm is due to the success of technology that has convinced most editors that this is the prevailing wisdom everywhere, including amongst academics or 'intellectuals'. Luckily, though, there is the internet, which may bypass the old-fashioned dependence of editors on ultra-conservative pundits in society. The editors of such media outlets usually have no familiarity with the mind/brain debate and thus are blissfully unaware of the idea of subjective consciousness and the debate surrounding it. If they were aware of it they might realize that the jury is definitely still out as to whether the physicalist thesis is correct or not. Because it is by no means obvious that the mind of subjective awareness is purely the result of physical processes, as opposed to digestion which is indeed the result of processes in the intestines.

So, contrary to the prevailing wisdom of the conservative pundits, it is altogether possible that reality has at least a dual nature as maintained by the dualist school of philosophy. There are several forms of dualism, however. Perhaps the 'tamest' of these is that favored by Chalmers, i.e. 'naturalistic dualism', which is the most materialist form in that it sees no influence on brain or mind outside of old-fashioned physical forces. Less tame is "interactionist dualism". Chalmers (1996) (P. 162) defines the latter as follows: "This view accepts that consciousness is non-physical, but denies that the physical world is causally closed, so that consciousness can play an autonomous causal role". Chalmers rules this out as he finds that current physics allows no loopholes for a non-physical mind to act through. However, as we shall see later, this view is more a bias than a cool assessment of the evidence. There is in fact rather a lot of evidence for such loopholes – modern physics is after all just a collection of rules describing the average behavior of matter. One example where the predictions of physics were overthrown was the quantum thing. There may be others, which we will discuss later.

Somewhat harder for most people to swallow except maybe for mystics in Hinduism or quantum physicists, is the idealistic school of philosophy, whereby matter is just illusion or 'maya' in the Hindu vocabulary. For quantum physicists, as the great physicist Arthur Eddington remarked in 1925, when Quantum Mechanics was still new: "it is difficult for the matter-of-fact physicist to accept the view that the substratum of everything is of mental character". As we shall see, a handful of physicists are prepared to face up to this interpretation of the weird world of the Quantum.

In the chapters which follow I will present some of the evidence from neuroscience, genetics and elsewhere that physicalist philosophy is by no means as triumphant as the majority of analysts of the press would have us believe. Thus although the genes specify much of what goes into the 'candle' of the body, they say nothing of the mysterious flame of subjective consciousness that lights the candle.

The great Humpty-Dumpty disaster in Neuroscience

As we saw in the first chapter, reductionist science has a few problems explaining some aspects of quantum physics such as 'mysterious action at a distance'. In biology also we begin to see the limits of reductionism when we come to such complex systems as the human body but even more so the human brain/mind.

The Grandmother Cell and Top Down Approaches.

As Malik (2001) says, the normal bottom up approaches of reductionist science such as are used in chemistry and physics will not work when applied to subjects, or subjective consciousness. Here, we need a top down approach, as the subject is an indivisible entity. Even in the apparently simple process of seeing, the old bottom up approach began to break down during the 'decade of the brain' (the 1990s). An example often used is the 'Grandmother cell' - a bottom up approach said that to recognize your grandmother's face, the image first forms on the retina and then is converted to nerve signals which are then fed to the brain where successively refined pattern recognition processes channel the impulses to a very small region of the brain, even as far as boiling down to a single neuron, the 'Grandmother cell'.

What neuroscientists actually discovered, however, was in a way reminiscent of the 'great ultra-violet disaster' in physics a century before that, which led to the invention of Quantum Mechanics (see the first chapter). One might call the brain equivalent the 'great Grandmother Cell disaster', as in a sense precisely the opposite happened to what was expected by simple reductionist bottom up theories - after some preliminary processing, the image is split up into its components: In one brain area horizontal lines are processed, in another color, in another motion etc. Part of the evidence for this comes from patients with brain injuries who can no longer process motion etc. Those reports usually focus on the negative aspects of such injuries. What they fail to mention, though, is the amazing aspect of the brain when everything functions perfectly: The granny's face is processed in modules all over the brain (as many as 40, according to Baars (1997)), with no idea of how to put Humpty Dumpty together so as to bind all the elements of the image processing again into a clear image - so we might refer to the 'great Humpty Dumpty disaster'. Note that there were anyway major practical problems associated with having a single cell for a view of every face, because you need many cells per face, to cope with all the possible viewing angles, which change the appearance of the face considerably. This would require an exponentially increasing number of cells according to how many faces one had to memorize. The argument might be alleviated somewhat by considering 'cell assemblies', or combinations of different cells for different memories. But although in this way it is estimated that the brain should have even more capacity than in the case where only single cells are considered, it is unlikely that all the permutations could be covered.

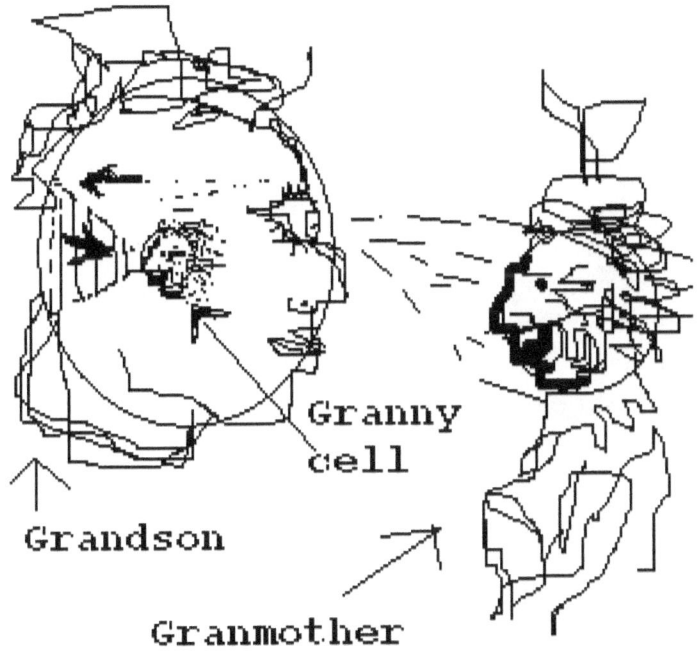

Figure 14: The old bottom-up view was that a single brain cell would recognize the grandmother's face. (Image by author ☺).

The 'Binding Problem'

This 'great Grandmother Cell disaster' is associated with one of the modern problems in the study of mind or consciousness, namely the 'binding problem'. That is, how do the data from these 40 or so distributed brain modules come together again to form a coherent image in the mind? How does the Humpty Dumpty of the different image features knit together into a seamless whole on our internal TV screen? Note that even if there were a single Grandmother Cell or Grandmother Cell Assembly, this would not explain how we have that internal TV image of the face on shoulders with a shawl and hands knitting a scarf on a rocking chair against flowery wall-paper. And not only vision is involved. Consciousness is a multi-media phenomenon, involving hearing, touch, smell and the body-sense as well as vision. Yet all these sensory data, whose processing is truly widely distributed in the brain, bind together in an amazingly coherent manner to give that multi-media experience that is subjective consciousness. Of course, Dennett & co. like to point to certain tricks used to flesh out this multi-media experience – e.g. we only have a small central region of the visual field which is well focused, and more peripheral objects are only hazily sketched in. However, this is merely nit-picking and says nothing about how those data which are present are bound together in that internal space that is our private virtual reality.

To the perceptual binding one must add the background thoughts that are the hallmark of consciousness. The latter include the internal conversation that is constantly on-going. Baars challenges the reader to go for more than 5 seconds without hearing at some level that internal voice commenting and planning. I took him on and only when I tried an old Zen-like form of ego-eliminating meditation did I break the 5 second barrier – without so much as an 'om'. But it was difficult – try it yourself. Also, occasionally we are thinking abstract thoughts. These must be added to the other 'contents of consciousness' to give an enhanced multi-media phenomenon. Those who seek evidence of quantum processes in the brain see in this binding of conscious experience a suggestive analogy with quantum non-local processes. Recall that this non-locality can mean that particles once part of the same quantum will stay connected though one goes to Alpha Centauri and the other comes here. Quantum non-locality is the only truly non-local process in physics – otherwise in the classical picture everything is disconnected particles. Whether there is literally a quantum component to the unity of the mental process or an analogous 'mental field' of some other sort is at present anybody's guess. But certainly something above and beyond the old classical separation and objectivity is needed to explain the unity or binding of consciousness.

The Being Within – the Self, 'Homunculus', or the 'Cartesian Theatre'

The hard core mechanist/physicalist school thinks with horror of the implication of this perceptual binding and the associated internal TV screen; namely of a little woman or man within the brain watching the internal TV screen: the so-called homunculus. The physicalists' alternative is to deny that there is any individual such as this homunculus watching an internal virtual reality show: they maintain that this subjective awareness of the homunculus, and that poor being herself, are illusions that are mere by-products of machine-like or purely objective processes. Amazingly, this counter-intuitive denial is absolute gospel to many neuroscientists (though not all, by any means), philosophers and others who are convinced that ultra-materialism is the only valid paradigm. But to the majority it is counter-intuitive to deny this inner self, soul, homunculus or whatever. The idea of such an essence has always been with us, ever since shamans danced round stone-age fires.

Despite energetic attempts to 'exorcise' the idea of 'self' or 'homunculus', it has refused to go away. See for example Sutherland's (1997) discussion on 'Homuphobia' or the fear of the homunculus. In this essay, Sutherland criticizes in particular the efforts of Thomas Metzinger, a German physicalist philosopher:

"Metzinger devotes just *half a sentence* (and tucked away in a footnote at that!) to the possibility that top-down effects may exist "in some cases" and then, panicking that the ghost of agency has not been truly exorcised adds: 'However, one *must* [Sutherland's emphasis] not conceive of such downward-processes as if there was a homunculus in the system, a little man pointing a beam of its already given awareness at inner states and thereby turning them into intentional objects: the phenomenal self, the centre of our inner experiential space, *must* [my emph] itself be thought of as a naturally emerged representational object, a transient computational model, which the system uses in organizing its behavior'. (Metzinger (1995), p 435, f.)

Sorry Thomas, the cat's out of the bag -- the magician has pulled the rabbit out of his hat. Just like the behaviorists with their "fractional antedating goal responses" (article Rosch, 1994), the agent has crept back into the system just like he always will."

Call it the 'Agent', 'Homunculus', 'Mind', 'Soul' or any of the words that strike fear into the physicalist, it means much the same thing. In his essay, Sutherland makes the point, also made by others, about the '40 hertz theory' for the binding problem:

"Hardcastle is quick to point out that the evidence for the 40hz. model as a solution to the binding problem doesn't really add up and it's interesting to note that much of the enthusiasm for this model has evaporated recently. The obvious question that one would ask is whether the oscillations are the *cause* or the *consequence* of perceptual binding. Is it just a spurious correlation? Put more directly:

how does the brain "know" that the output of certain feature-detecting neurons is part of any particular object without a prior (top-down) model of that object?"

The tendency for researchers in the field of consciousness studies to treat any process that seems to be synchronized across the brain, such as the 40 hz oscillations, as a panacea for the problems of subjective consciousness, is indeed remarkable, since, as indicated above, the mere fact that several neurons associated with processing an image are oscillating in sync does not explain how I see that unified image on the internal TV screen. Like the quantum tendency to associate quantum non-locality with conscious non-locality, all we have is a correlation, not an explanation. Thus the 40 hz may or may not be one of the NCC, or 'neural correlates of Consciousness'. But even if it is an NCC, this is related merely to the 'easy' problem of consciousness and not the 'hard' one of how the 'wine of subjective experience' emerges from the water of the NCC. Those who focus on the 40 hz 'explanation' of consciousness at some point in their description need to write 'and then a miracle occurs'.

The need for a top down approach in neuroscience and related fields is acknowledged by a growing number of neuroscientists. As mentioned in McCrone (1997):

> "Traditional thinking held that the brain was some kind of computer, crunching its way through billions of inputs each second, outputting consciousness. But, said Friston, a theoretical neurobiologist at London's Institute of Neurology, it is more as if the arrival of those inputs provokes a widespread disturbance in the brain. Look, Friston told his Harvard friend, the brain is like this pond. You throw in a pebble - the sensory input - and you get ripples. That's the neurons responding. Sure, the pattern says something about the way the pebble hit the surface. But the pond is already

covered in ripples caused by other pebbles, so the pattern appears a little chaotic. And then once the ripples spread out far enough to begin bouncing off the sides, he continued, the shape of the pond begins to affect what is going on. The whole thing keeps evolving and becoming more complex. Yes, replied his friend, nodding furiously, and as we throw more and more pebbles - or rather experiences - into the pond, we change the kind of patterns it produces, and even the shape of the pond itself. This system has a memory! In the early 90s, in hundreds of private conversations like this, mind scientists were groping their way towards a fresh view of the brain - one based on the idea that mental states are dynamically evolved rather than clinically computed. Back then, the arguments were little more than hand waving exercises. People were familiar with the new ideas about chaos, complexity and non- linear systems coming out of places like the Santa Fe Institute, but unsure how they applied to the brain. Today, however, the dynamic revolution is beginning to roll. At workshops and meetings around the world, researchers like Friston are talking publicly about dynamic models of the brain, and the evidence to support the new theories that is beginning to fall into place. A replacement for the brain-as-computer model certainly seems overdue."

So there is a tendency amongst 'mind scientists' to acknowledge top-down process in the brain. Another piece of support for that was the research mentioned in Chapter 1 on monkeys that showed that there was feedback all the way to individual neurons on the retina if those neurons were part of an image of a banana or other object of interest. And speaking of single cells and feedback – in bio-feedback experiments test subjects have been able to control the output from individual nerve cells on their backs simply by focusing on the output of its activity on an oscilloscope and trying to selectively change it. The latter is indeed remarkable, as we would normally have no need of such individual cell control. Yet all we have to do is direct consciousness at the task and it is accomplished.

But though many scientists now accept top down processes and feedback as part of the brain/mind system, few make the connection to the homunculus. Yet it is ultimately the only logical conclusion. Every physicalist comes to some point in her description of neural processes where it is declared that the magic process has been found which really ' owns' the other processes, just as a self should. For many it is the 40 hz process. For others it is something else. E.g. the famous neuroscientist Ramachandran says that he wishes to 'approach the greatest scientific and philosophical riddle of all – the nature of the self', but suggests that the audience for perceptual processes is not a self but other brain processes – this time in the brain's limbic system. This is the point in his argument where 'a miracle occurs'.

Also with neuroscientist Antonio Damasio, in his multi-level scheme, the self is transient, constantly recreated for each new object with which the brain interacts. There is an internal movie associated with autobiographical memory and Damasio says that the autobiographical self is both the owner of the movie and somehow emerges from the movie. But his is just another sleight of hand in shifting the emphasis from process to process without ever saying what subjective consciousness is or how it is associated with these objective processes.

Many philosophers of mind have realized that the only logical alternative to the 'little woman in the brain' is 'infinite regress' or to maintain that consciousness is an illusion. By 'infinite regress' is meant that if Mary II is the one looking at the internal TV screen of Mary I, then there must be a Mary III looking at Mary II's screen, ad infinitum. That appears even more ludicrous than the Homunculus. As for dismissing all human experience as illusion – as Descartes said in his famous treatise where he concluded 'cogito, ergo sum' ('I think, therefore I am'), when he had successively eliminated everything of which he could not be certain, using his famous power of logic, the only thing left was that inner core of subjective consciousness.

Thus in cognitive science or consciousness studies one often hears of the 'Cartesian Theatre', where the images are cast on a screen or stage by a team of backstage (sub-conscious) workers. Baars (1997) is one cognitive scientist who finds the idea useful as a metaphor, as it explains so many features of what goes on in a model of mental activity. The point is that subjective consciousness is the 'place' where the end-result of all the subconscious processes, modules, or backstage workers becomes visible and is broadcast or made generally available all over the brain/mind. This in turn suggests the idea of the audience. But of course this is another way of saying there is a little person or persons in the brain looking at an internal screen or stage. Thus the homunculus and the Cartesian Theatre are more or less the same thing. Since the 'metaphor' fits the description of mental processes so well, I am suggesting that Baars' metaphorical theatre is in fact how it really is.

So for physicalists like Metzinger or Dennett to say 'I see processes going on in your brain when you're thinking and so your self is an illusion' is rather illogical, to say the least. In other words, they maintain that observations by one *subjective observer* of objective brain process correlates of the *subjective awareness* of another person indicate the illusory nature of both of them! Descartes realized this with his impeccable logic, and indeed it is 'as plain as the nose on your face' that your awareness is the only thing you can really be sure of. To see just how shallow is the argument of physicalists, let us look a bit more closely at views in this direction.

Critique of the Behaviorist or 'Physicalist' view – for whom is the illusion?

In his book 'Consciousness Explained', Daniel Dennett skillfully presents his viewpoint using his method of telling stories to illustrate what he takes to be obvious features of consciousness. His book was unusually successful at the time it was published. His success with general readers lies in this ability to weave a story around his theories. Somehow, though, he manages to pull some rather curious rabbits out of his hat, and persuades the reader that they are somehow relevant to the argument as to whether subjective consciousness really exists or is a subtle illusion of brain processes.

But, as Churchland and Ramachandran point out in the conclusion to *their* critique of Dennett, presented in the book 'Dennett and his Critics' (Dahlbom 1993), there is for example 'nothing whatever (no fact of the matter) to distinguish between a misperception and a misrecollection'. Yet Dennett goes to great length in 'Consciousness Explained' (call it CE from here on) to distinguish between these two processes whose nature is dubious from a scientific perspective. Churchland and Ramachandran also correctly point out that his devotion of a large part of CE to the way the brain 'fills in' for missing (perceptual) data, e.g. for the blind spot of the eye's retina, is not very relevant for dismissing Cartesian theatres. They correctly mention that such processes may occur rather early in the subconscious processing of the data. It is indeed one of the points of a Cartesian theatre that there is much activity behind the scenes before the finished product is presented to the 'audience'. As implied by Churchland and Ramachandran, all the discussion of the pros and cons of Stalinesque, Orwellian or Multiple Drafts mechanisms by Dennett are just a screen for a rather behaviorist approach. Though Dennett himself would deny any precise definition as a behaviorist or an instrumentalist, it is plain that he leans in that direction. I.e. just as the behaviorists ignored everything that had to do with subjective consciousness, so too does Dennett seek to dismiss as an illusion the idea, e.g., of the fact that one actually has a subjective awareness of seeing a red flag (speaking of Stalin). That is, the 'qualia' are simply illusions. As Churchland and Ramachandran put it:

'Simplified, the heart of Dennett's behaviorism is this: the conceptual framework of the mental does not denote anything real in the brain. The importance of the framework derives not from its description of neural or any other reality; rather, it is an organizing instrument that allows us to do fairly well in explaining and predicting one another's behavior, the literal unreality of qualia etc. notwithstanding. How is it that the framework manages to be a useful instrument, despite the unreality of its categories? Because, according to Dennett, even though there is nothing *really* in the brain that corresponds to visual awareness of red, there is *something or other* in the brain which, luckily enough, allows us to get on pretty well in making sense of people's behavior on the pretense, as it were, that the brain really does have states corresponding to awareness of red.'

Their italics are just as I would have chosen, as this is indeed the crux, where Dennett's slight of hand is revealed. Just as with Sutherland spotting Metzinger's sleight of hand in his footnote (see above), there is, as Sutherland says, always a point, despite reams of distracting discussions of peripheral phenomena, where the cat is let out the bag and 'the agent has crept back into the system just like he always will.' One might as well have written 'and here a miracle occurs'. I.e. for a behaviorist, a person is just a behavior machine; according to this idea, the 'sensori-motor' behaviorist set-up allows us to react with our motor system (muscles etc.) to the sensory data. But for the behaviorist there is nothing between sensing and acting except a complex bridge of brain processes between the two which does the job of recognizing the tiger and developing a strategy to deal with the situation (run like hell). Similarly, Dennett posits *something or other* that takes the place of the actual visual awareness of red, but has the same effect. This sounds a lot like the behaviorist 'zombie' solution. It is his sleight of hand – some mysterious 'something' must do the same job as visual awareness. He implies that we are only fooling ourselves when we claim to see red – it is just an illusion of the sensor-motor loop. But an illusion to whom? An illusion must have an audience to be appreciated as such, as must a perception.

Again many, steeped in the lore of the brain lesions and deficits in the popular literature a la Dennett, think that arguments such as, 'Synesthesia (where some people e.g. see colours when they hear certain words or music) appears to be an example of illusion - a way to fool the brain' might be relevant to the topic. However, though the brain/mind is indeed confused by the mixing of sensory modes involved in Synesthesia, it doesn't necessarily follow from this that all is illusion: The mind knows perfectly well that a 'feature' has been added to its perception of words, music or numbers. But again this is consistent with the end-user, the internal experiencer, Homunculus or whatever, noting the finished product that is projected on its internal virtual reality – the fact that it differs from the normal system does not imply that the recipient of these sensory 'finished products' is an illusion any more than the experiencer with normal brain systems.

Again coming back to Descartes: his point was that our subjective awareness is the only thing of which we may really be certain. And it is so much more natural to accept the fact that we really are seeing red and not that an illusion for nobody is occurring! The feel of the experience of red is unique and, to use a much hackneyed expression in the field, ineffable. There is indeed somewhere a screen on which your field of visual awareness is projected – only it is not, as McGinn (1999) points out, located in the brain. It has no location, but occupies a sort of phenomenal or subjective space. We build up a form of virtual reality parallel to 'the thing in itself', as Kant might have said. Also, our abstract thoughts do not occupy any real space but float around in an alternative, inner universe.

So once again, it is irrelevant that our eyes jump around the place in little 'saccades', a motion that is filtered out of our visual awareness. It is also irrelevant that we have just a narrow range of acute vision and an inaccurate peripheral vision. The net effect is that it constructs this inner virtual reality which is perfectly adequate for all our activities, be it walking up a staircase, driving a car or playing tennis. The agent will always creep back into the system to become the viewer in the Cartesian theatre. And that agent is the dreaded Homunculus.

Thus although I have quoted Churchland and Ramachandran above in their criticism of Dennett, I profoundly disagree with them when they say that 'Dennett brilliantly and quite properly debunks the idea that the brain contains a Cartesian Theater wherein images and the like are displayed' or "… some categories such as "the will" and "the soul" probably do not correspond to anything real…". On the contrary, the Cartesian theatre and the homunculus are still very much alive. It is the behaviorist or physicalist illusion that is in need of debunking, as we shall see in the next section.

Confusion amongst the physicalists

Susan Blackmore is another thinker in the field, who in her 'Consciousness – a very short introduction' sides herself with Dennett in his attempt to prove that consciousness, qualia etc are just illusions. Her description of the Dennett or physicalist viewpoint highlights some of the weaknesses and contradictions in that school of thought. For instance, she asks how consciousness can be effective, i.e. useful in any way. To show that it cannot, she asks how just qualia or subjective awareness of red roses, cats on sofas, etc. can confer power on the mind to influence matter. The obvious answer here is, as pointed out by Baars etc., the 'sensori-motor' control loop: If I see a glass and wish to grasp it, the input image is used to guide the mind in adjusting the motor responses of the arm and fingers to grasp the glass. Thus mind is in the loop for controlling actions via sensory feedback. That could also apply to a program in a computer for pattern recognition of glass and hand, with use of a 3-d visualization system to predict actions and issue the suitable motor commands. This is indeed the case for the most advanced high-performance robots.

But again the robot is just a model of the mind-body and it is by no means certain that it is even close in its model. For one, complex actions like grasping a glass still have to be helped along by human programmers with programs to guide the robot in the right direction – though the robot may have a degree of autonomy in neural networks that allow it to 'learn'. But in state of the art models, the robot's action is still guided to a large extent by linear programming, whereas we know that the mind is massively parallel in its architecture. The fact that qualia are in the loop of the human control cycle implies they are indeed part of the process and thus DO lend power to the mind to interact with matter. Baars, Llinas and others go further and say that consciousness was a necessary survival skill throughout evolution. If a tiger is charging at you, dry zombie-like action is not always appropriate. Better to have a subjective mind whipped into panic by adrenaline that sends shock waves of painful intensity to the perceiver within, who then has a stake in developing a strategy for survival – e.g. "get the hell out it, like yesterday!"

Another weakness of the Dennett / physicalist standpoint is a tendency to confuse various issues. For instance, denying that there is a place in the brain where it 'all comes together' is taken as disproving the existence of a mind or subjective consciousness or internal TV-screen. However, this is a straw target, as the dualist point of view does not necessarily require a central processor in the brain or a position in physical space where it all comes together. The modern form of Cartesian dualism does not seek out the pineal gland or some central processor in the frontal cortex to link brain and mind. Since the brain processes corresponding to almost every mental experience have been shown by brain scans to be widely distributed, it is apparent that any link from mind to brain or subjective consciousness to neural processes has to occur over a wide area if not all of the brain. Thus if there is another dimension of subjective awareness, as Colin McGinn implies, a sort of virtual reality corresponding to inner dimensions, it is apparent that this inner space can map onto brain processes at many points to form a distributed correspondence between brain and mind. Also, the mystery of perceptual binding indicates that this may indeed be the case – a visual image is dissected into 30 or more specialized modules – but comes together on an internal screen – or by feeding into an internal virtual reality for the delectation of the perceiver, soul or homunculus. Thus the imperative need of the physicalist to be in denial of the homunculus leads to a curious and erroneous conflation of ideas.

Another of these misguided associations of Dennett and co. is the idea that because consciousness is not as smooth as William James thought it was 100 years ago, then there is no smooth subjective awareness. But again, it is irrelevant if vision is interrupted by 'saccades' or jerky movements and the blind spot, which are 'smoothed over'. Also, 'change blindness', where we are unaware of details of a scene which change when we blink, says nothing of the qualitative or holistic sense of the awareness. For the same reason, though wine never tastes exactly the same, that does not show that we never taste wine or enjoy it – those of us who are not teetotalers. Again the physicalist are beating a straw man when they attack quantities of the picture – a pixel here, a pixel there etc. without understanding that subjective consciousness is about the holistic whole or quality of the experience. I.e. it's about the inside, not the outside of the experience. As Wilber and De Quincey (De Quincey 2002) repeatedly emphasize, every outside has an inside. Dennett & co. focus exclusively on the outside, and deny the inside. But we know the world only from the inside. As De Quincey's stance is that of a pan-psychist, he sees consciousness as going all the way down to atoms, in rudimentary form, i.e. there are outsides and insides all the way through from atoms to humans. But even if we don't accept this radical extension, his arguments are still good applied to humans alone.

The latter point should be emphasized, as physicalists repeatedly return to the imperfections of perceptual awareness to maintain that it is all an illusion. Thus Blackmore makes much of the fact that we only have sharp visual awareness in a very small region at the centre of the field of vision, corresponding to the Fovea region of the retina. This, to her, is proof that visual awareness is an illusion, as we are not really aware of the whole room at one time. We only see the computer screen in front of us clearly – however, we do have a sketchy awareness of the periphery. E.g. right now I see the keyboard or screen of my PC but to the left I am aware of clothes on hangers in a wardrobe and paintings on the wall and beyond that one window while on the other side I am aware of the TV screen and the other window etc. Yes, these peripheral objects are not in sharp focus, but I have a rough model of them and to refresh my knowledge of them I can glance over to view them in detail. To maintain that this system is tantamount to there being no vision at all is ludicrous, simply because awareness of the way it is built up normally hides the fact that we see mainly with this central sharp focus. I just had another example to emphasize this. A blackbird flew from a branch in my garden: first I just noticed a blur of motion to my right, then I swiveled my gaze quickly enough to catch the bird in flight before he left the garden. This is typical of the flowing synergy of peripheral and central vision. Also, it's not as if there was a sharp cut-off with tunnel vision ahead and all else a blur: the fade off is gradual, as in most biological systems. For the same reason, though wine never tastes exactly the same, that does not show that we never taste wine or enjoy it – those of us who are not teetotalers.

Just leafing through Blackmore's 'Consciousness – a very short introduction', I again spotted a typical example of her sleight of hand in making consciousness disappear: Simply by stating that science does not need the brain to have an owner, she concludes that there is no 'need' for an 'inner experiencer' (p. 66). This assumes a lot about the nature of science and its brief. In fact, the original meaning of science was very broad. The word science comes from the Latin for 'to know'. Thus if there is an inner experiencer, the brief of science should be to get to know it. Hence the old exhortation of the Greek philosophers: 'Know thyself'. Descartes was quite correct in his wonderfully clear analysis that logically eliminated one after the other all aspects of reality which might be illusions – what was left was his subjective consciousness – 'I think therefore I am'. Thus the stance of physicalists like Blackmore and Dennett in declaring consciousness an illusion is precisely the inverse of the logic applied by Descartes to the nature of his existence. The one thing which is not an illusion in Descartes' analysis is the one thing that IS an illusion in theirs. When they appeal to Science to support them in this stance, they are thinking of just one school of thought among some scientists which has taken on board the philosophical baggage of physicalism and thus declares ab initio that only that which fits into a sort of Newtonian objectivity has any reality. But Newton has been surpassed by Planck and Einstein, and as a true scientist remains 'objective', it is ironically by remaining objective about the subjective that a scientist remains 'pure' and not a corrupt cop knuckling under to the mafia that will make life hard for him if he doesn't play ball and deny any reality to the subjective (sorry – I was just watching '16 Blocks', where Bruce Willis defends a key witness from corrupt cops).

We saw in the first chapter that quantum mechanics implies that reality may be taken to be completely mental, with no hard and fast objects in the old Newtonian sense.

As Richard Conn Henry indicated in his paper in the American Journal of Physics in 1990 'Quantum Mechanics made transparent', the quantum revolution was more profound than the Copernican one. He agrees with Arthur Stanley Eddington when in 1925, shortly after the discovery of Quantum Mechanics, he stated (as I already mentioned earlier in this chapter) that "it is difficult for the matter-of-fact physicist to accept the view that the substratum of everything is of mental character". Dennett and Blackmore, perhaps because they lack a physics background, fail to appreciate this revolutionary shift in world-view and believe those (usually chemists and biologists) who state that quantum effects may be banished to insignificance when dealing with macro-systems.

So die-hard behaviorists and physicalists need to grasp at straws such as the imperfections of the perceptual and active memory systems. Chalmers (1996) makes it clear how the crack in their argument is papered over when ultra-materialists or physicalists maintain they have explained away consciousnesses. He points out that every mental process has what he terms a psychological or objective side, as well as a phenomenal or subjective side. By homing in on one of the psychological processes such as visual perception, or introspection, or reporting on one's thoughts (reportability), such writers often claim that by providing a plausible physical mechanism for the psychological process, they have explained away its phenomenal side as well. Quite the contrary, as Chalmers points out. Just as the conjurer's complex actions conceal how he hides the trick, these writers seek to obfuscate by talking at length about these psychological, objective processes, and ignoring the subjective or phenomenal side of the process. They are attempting to explain what Chalmers calls the 'easy' side of the problem of consciousness, whilst saying nothing about the truly hard problem of how the subjective arises from the objective.

Antonio Damasio is a hero of the physicalist camp, and his analysis of how emotions flavor our thinking, e.g. in his book 'The Feeling of what Happens' is often cited as another element in exposing the illusion of consciousness. But feelings are nothing other than examples of qualia – their more important side, that is, i.e. the subjective side. True, there are objective correlates. But the driving power of emotion is how its subjective urgency can influence and 'hijack' the central spotlight of awareness and drive the subjective will in different directions.

Chalmers' analysis is also relevant to the debate about free will. The usual way to deny free will is to explain how some objective subconscious processes (i.e. with no subjective conscious counterpart) seem to build up before the subjective awareness of a decision has been made. As we shall discuss later, "free will" need not be an illusion, no more than "the self" or subjective consciousness.

The Arrow of Time and Subjective Consciousness

Another mysterious aspect of consciousness is its subjective sense of the flow of time. Most popular discussions on time are unsatisfying in getting to the nub of the matter. Usually all that is discussed is Einstein's relativity and how it can expand and contract perceived time for an observer moving relative to another. But these discussions, though they show how the subjective sense of time may be distorted in the measurement of objective entities such as a meter rule or the tick spacing of a clock, say nothing of the essentially subjective perception of the flow of time. Shallow dismissals of time as an illusion, often beloved of ultra-mechanistic physicists (e.g. Barbour (1999), Hawking (1998), also fail to come to terms with time as perceived. Some physicists, however, recognize that time must be given a very special treatment and that one of the problems with recent attempts to construct a 'theory of everything' is that there is something missing in the way time is handled in the equations. Lee Smolin (2006) is one physicist who has emphasized that even the cutting edge of physics has failed to tackle the problem in a meaningful way.

The subjective sense of Time

No amount of trying to sweep it under an objective carpet can provide an insight to time as perceived by subjective consciousness. In fact the two are intimately related: without the subject, time would boil down to the trivial objective time of the physicists: just another coordinate axis plotted against a space axis. In relativity time is objectively equivalent to a space coordinate multiplied by the square root of -1. Yet subjective time is implicit everywhere, even in perceiving a picture or a graph of space versus time: The physicist could not look at that graph if there were no perceptual binding of the image, which also takes subjective perception processes and subjective time sense to form. Philosophers such as Husserl and Heidegger (1997) do make tentative inroads on the mystery of subjective time, but basically get no further than describing a sort of extended `now` out of which flows the past and into which flows the future.

In a German issue of a History of Philosophy by Hans Joachim Stoerig, in the section on Heidegger, I finally grasped a few things about the great philosopher (author of the famous 'Being and Time') better than before:

* His philosophy, just as Husserl's and the other phenomenologists', is rigorous but not in the same way as science. Thus he is distinct from and in contrast to the rather unpleasant and boring logical positivists.
* It is interesting that he applauded Hegel for at least

one of his ideas - namely that absolute nothingness is actually the same as absolute being. Heidegger's take on that is that nothingness is the veil of being - a bit like chiaroscuro, shadows and light.

In this sense being is transcendent in that it stands out from the void and is more than just a 'place marker in nothingness'.

* He makes the point that philosophy is actually closer to poetry than science. Good point - as 'science' in the sense of objective science demands that we be zombies, whilst poetry is exclusive to subjects. The famous example of a sunset - objective scientific description is trivial, whilst a poet gives a lyrical account of subjective qualia. - this point struck me again in reading also an interview of an editor of Spiegel with Brian Greene of 'Elegant Universe' - in his new book he again shows how one sided he is - apparently utterly ignorant of philosophy, in the interview he said things like:

"Oh the flow of time must be an illusion as it only occurs in subjective accounts".

This is unbelievably ignorant of a supposedly intelligent scientist, that he should have read no philosophy dealing with this essential aspect of existence - all existence is subjective, in that it is perceived, some would say created, by subjective consciousness! Of course, when pressed to say how time was created before the big bang, Greene had to confess ignorance. He also had no answer as to why the time dimension should appear so different from the spatial ones, although mathematically they are on a similar level. To quote from http://observer.thecentre.centennialcollege.ca/features/briangreene.htm :

"But at high speeds those watches can be off by seconds, minutes, even years if
they move fast enough. Therefore, the whole notion of past, present and
future is nonsensical and completely subjective. "

In the latter utterance once again this philosophically challenged innocent thinks that just because it's subjective it's nonsensical - the opposite is the case! He also shows his failure to grasp the concept of qualia - it is immaterial if the subjective impression of time should be stretched or compressed a bit by relativistic effects - this says nothing about the *quality* of what is being stretched - namely the subjective impression of the flow of time.

None of the popular discussions on time are satisfying in getting to the nub of the matter. Usually all that is discussed is Einstein's relativity and how it can expand and contract perceived time for an observer moving relative to another. Quantum theory comes a bit closer to the mark with its discussion of the collapse of the wave function. As Henry Stapp (1993) notes, these wave function collapses are irreversible and so point to a preferred direction or arrow of time. This is more general that the usual idea of the thermodynamic arrow of time, as the latter is only true statistically for large numbers of molecules, while the quantum collapses are true at the smallest of scales. Still, though there is a mystery about such quantum processes, there is no guarantee that it is the same mystery as that of subjective consciousness. Subjective awareness of a landscape or anything else, including a graph of x versus t, is a complex interplay of non-local perceptual binding in the brain and a subjective sense of Newton's equable flow of time.

One friend phrased a query typical of the new breed of computer science physicalists (geek materialists): "Can't subjective time be sufficiently explained using memory content? I have the feeling that yesterday is before tomorrow because I have memories of yesterday and none of tomorrow. This generates the whole (to the physicist spurious, and rightly so, say I) illusion of an advancing edge of present at which future becomes past. A creature with no memory can have no sense of time passing and, setting aside the slight impediment of comprehensive amnesia, would probably make a better physicist than you or me."

Memory content a la computer data storage ('working memory') is just an objective correlate of the subjective impression of the recent past fading away gradually, while we have a stronger subjective awareness of the present. Similarly, we anticipate the near future in planning, but it is not yet present in subjective time awareness to the same extent as the 'now'. This is really like a Zen Buddhist 'koan' - how time's arrow is subjectively felt. No words can express it, just as words fail when describing red or C minor. They are ineffable. Philosophers before Descartes were caught up in the adulation of rationalism as true thought, and often failed to see the paradox of subjective experience. The 'koan' of the sound of one hand clapping is nothing compared to the koan of the subjective impression of red of the scorched palm. And for time, the case was well stated for the 'time koan' by Saint Augustine, who in the 4th century already stated, in his 'Confessions', the basic paradox of the subjective nature of the arrow of time:

"What, then, is time? If no one asks me, I know what it is. If I wish to explain it to him who asks, I do not know."

The subjective nature of the arrow of time is not addressed by thermodynamic arguments such as those of Hawking or the recently deceased Ilya Prigogine. Such objective processes don't get the point of the arrow - which is again subjective and a direct corollary of the qualia of the time sense. Husserl describes the simultaneous holding of different points in a time interval in consciousness in his analysis of the sense of music - we would have no feeling of music if we were only ever aware of one note at a time. Somehow the relation between past and present notes excites us - in this sense music is like a subjective enjoyment of maths, since the relation of notes and intervals to each other is essentially mathematical. And music is only subjective - an objective analysis is dry as dust.

So the sense of time is tied up with the binding problem. What aspect of mind allows it to bind disparate processes scattered all over the brain and integrate them into a coherent sense of perception – vision, hearing, heeling, and all that moving through the world along the time axis? This is the masterpiece of consciousness – integrating and binding within the rhythm of time's passage. There is indeed something missing in candidate theories of 'everything' in physics if they cannot describe this time-binding process. Perhaps for this reason have some of the best mathematical thinkers of the past century turned to the philosophy of consciousness. Notable amongst these was Alfred North Whitehead. In his book 'Process and Reality' (Whitehead 1978), he "saw the definite character of events as due to the "ingression" of timeless entities" (Encyclopaedia Britannica, 2006). This philosophy led to Process Theology. Two of the important tenets of this theological school are:

☐ The universe is characterized by process and change carried out by the agents of free will. Self-determination characterizes everything in the universe, not just human beings. God cannot totally control any series of events or any individual, but God influences the creaturely exercise of this universal free will by offering possibilities. To say it another way, God has a will in everything, but not everything that occurs is God's will.

☐ God contains the universe but is not identical with it (panentheism, not pantheism). Some also call this "theocosmocentrism" to emphasize that God has always been related to some world or another.

(From Wikipedia article http://en.wikipedia.org/wiki/Process_theology).

So we see that one of the most profound thinkers in mathematics of the 20th century also led to a school of thought in theology that sees the universe as evolving in a way that is consistent with free will. Thus the temporal process of evolution is part of God's plan (either so ordained or dictated at least in part by the nature of existence). Time then is just this process of becoming. Not surprising, then, that it is so hard to capture the flavor of actual time in an equation. It takes a great mathematician to realize this.

Although Hawking has been cited above as a physicalist, in one place at least he makes a salient point regarding the ineffability of time, space and existence, originally applied, I think, to the universe as a whole. Physics gives us, "no idea of what breathes fire into the equations and makes there a world for us to describe". Maybe music and the subjective time sense are good examples of 'breathing fire into the equations'.

Breathing fire into the Equations

In many modern theories of physics, there are a number of possible universes. This is the case in String or Brane theory, where many parallel branes may co-exist, each with a different universe such as ours. Then, as we saw in the introductory chapter, there is the many-worlds interpretation of quantum mechanics. But in the standard interpretation of quantum mechanics, called the 'Copenhagen interpretation', there is just one universe which evolves via successive 'choices' that particles make. So whilst the many worlds interpretation sees the outcome of each uncertain particle movement as all of the possibilities being realized by splitting off ever more copies of the universe where all these possibilities are fulfilled, the Copenhagen interpretation sees a random 'collapse' of the set of possibilities onto one 'winner'.

This selecting of the 'winning' possible universe amongst all the possible ones is one way in which the equations point to a process which makes the possibilities concrete. This process may be seen as a random one, or in some sense involving free will, if one can speak of particles as having free will. This might be the case if a sort of pan-psychism is true – i.e all matter is in some sense alive, even at the level of atoms or photons.

The other sense in which equations lead to something concrete is Hawking's breathing of fire into the equations, i.e. hauling them out of the abstract realm of mathematical objects into a real world of concrete objects. This could be one way in which God intervenes directly in the workings of the universe – by a continual miracle, the mere abstract equations of Einstein, Newton, Planck, Heisenberg etc. getting realized with real matter. And this constant miracle takes place within an equally miraculous thing we call time. As Hawking says, science can't tell us one way or the other whether this picture is the right one. It can only describe the magic formulae – but it needs a magician to speak the magic spells that conjure reality of no-space and no-time.

Free Will, the Self and the sense of the numinous – not just Illusions!

There are many ways to attack free will. One which was repeated in "The Science of Discworld II; The Globe" (Pratchett, Stewart and Cohen (2003)) was that of limited choice essentially forcing us down one or other path. On the positive side, however, the same book points out how quantum uncertainty and chaotic sensitivity to initial conditions indicate the falsity of pre-destination. Also, the same writers draw attention to the way in which creativity took off when homo sapiens sapiens (i.e. humans) learnt to tell stories - 'pan narrans' or the story-telling ape. I will argue here that the stance taken in that book is somewhat inconsistent. Also, I will provide arguments against the main weapon in modern neuro-reductionism's argument for free will being merely an illusion – i.e. the apparent lack of involvement of conscious will in the decision process as indicated by brain scans that show a build up of sub-conscious processes a fraction of a second before the person declares herself aware of having made the decision.

Evolution of the Human Brain, Mind and Will Power

Modern students of human evolution agree that the deciding factor in accelerating the human revolution (art, agriculture, technology) was the explosions in creativity of our ancestors when they split off from the main homo sapiens group between about 80,000 and 100,000 years ago and even more when artistic creativity emerged out of nowhere about 35,000 - 30,000 years ago. . The contrast is striking with groups such as the Neanderthals, who remained relatively stagnant for tens of thousands of years and while skilful in surviving the harsh ice age conditions of Europe at that time, were not very creative compared to modern humans. The implication here is that the sudden advent of creativity caused a relatively narrow set of possible futures to blossom into a many branched tree of possibilities. Thus creativity expanded the degrees of freedom of humanity enormously, in the same way that a gas with more heat energy expands. On might say that the free will space was extended enormously.

In the long history of the evolution of life on Earth, there were several notable points of acceleration – one was the Cambrian Explosion – the time, about 540 million years ago, when the few species of sea creatures present up to then 'suddenly' developed into a plethora of forms. In the space of a few million years, but a twinkling compared to the thousands of millions of years since Earth and life began, many strangely formed animals developed, as if nature were suddenly experimenting with forms. Now part of the answer to this increase in complexity of animal forms was the development of the Hox genes, which are present in the genomes of nearly all multicellular creatures. They are markers which give the rough structure of animals. Essentially the same Hox genes in insects are associated with the abdomen, thorax, head etc. as in mammals with head, body, legs etc. But note 'associated with' and not 'give', as the old simple picture of specific genes being alone responsible for some feature is no longer thought to be correct. Almost all features or sicknesses arise from the complex interaction of many genes – cases of a one to one relationship are extremely rare. Thus there are only a handful of diseases that may be easily treated with gene therapy. So the explosion of complexity in the Cambrian was in turn associated with a greater complexity in genetic interactions, of which the Hox genes were a part. One other possible feature of this Cambrian explosion may have been that subjective consciousness 'woke up' for the first time – i.e. a threshold of complexity was passed after which creatures acquired a crude sense of awareness (note: not the same as self-awareness).

The passing from the zombie phase to the conscious phase may have opened up as many if not more possibilities than the evolution of hox genes – indeed as feedback is so rampant within the body and it has even been shown that brain activity feeds back on the genetic structure of neurons (Candel, 2005 ****), then it is even possible that hox evolution was accelerated by the positive feedback action of consciousness: a rather speculative extension of Lamarckian evolution perhaps, but possible according to modern research in micro-neurobiology.

Free will is often challenged by writers in neuroscience and related areas of philosophy (e.g. Rita Carter, Tomas

Metzinger). Of particular interest here is Benjamin Libet's experiment which showed that for 500 milliseconds prior to a conscious action subconscious activity (e.g. commands to move an arm) is building up. Libet himself (e.g. Libet, 2003), however, does not interpret his experiment as evidence of the inefficacy of consciousness - he points out that although the tendency to press a button may be building up for 500 milliseconds, the conscious will retains a right to veto that action in the last few milliseconds, as has also been shown by experiments. A good comparison made is with a golfer, who may swing the club several times before striking the ball. Also, James (1890) viewed consciousness as a continuum, reaching from the peak of full directed waking consciousness down through layers of lower awareness: thus to say that 'subconscious' processes determine our actions could also be stated otherwise: The action simply starts at a lower level of awareness, coming to fruition only when it has risen up to the peak of conscious awareness. In this view, the action simply gets, as it were, a rubber stamp of approval at the last millisecond. So the thought was never fully disjoint from consciousness ('sub conscious'), but was always rising through successive layers of awareness. This picture reminds one perhaps of Dennett's multiple drafts – but a version of his theory could also have multiple subconscious drafts of which only one is selected by conscious free will. That in turn reminds one somewhat of Henry Stapp's ideas about quantum consciousness: alternative brain states corresponding to alternative lines of action may be present as a superposed ensemble of quantum states – ghostly probabilities which again conscious will collapses into one concrete choice.

Some dismiss Libet's 'get out clause' of the last minute veto (e.g. Brooks 2008). There are other arguments about the fuzziness of the moment of decision – that narrows the gap by a few more milliseconds. One nice set of tests that I find could provide a neat explanation of the backdated decision is that performed by Radin, Bierman et al. (Radin (2006)). I include the following description here instead of in the chapter on 'More Things in Heaven and Earth' because of its relevance to the Libet experiment. Radin and coworkers' performed experiments on 'presentiment', or the supposed ability of the mind-brain to sub-consciously sense up to a number of seconds into the future – i.e. subconscious reaction is measured by wires which record skin conductivity and other data usually recorded by lie-detectors. The test person, wired up like this, is then presented a series of images on a computer screen. These images may be soothing, neutral, or disturbing in being violent or sexually arousing. Radin and co. made sure that the database of pictures was large enough that no problems arose by re-using pictures etc. The results should then have been well balanced and so, if there was no effect due to presentiment, there should have been equal chances of the 'lie detector' reading increasing before a neutral or exciting image was revealed on the screen. On average, however, they found a rise in anxiety more often for the exciting image. See e.g. Radin's blog: http://deanradin.blogspot.com/2008/09/presentiment-demos.html . It's interesting that he usually gets good results on live TV shows, where conditions are not as well controlled as normal. The experiments were reproduced by Bierman and others and so cannot be dismissed so readily. This effect would balance the Libet effect, in that one could imagine the subconscious anticipating the conscious decision and already initiating steps to accomplish it. This would also tie in well with the last millisecond veto – since the conscious will would both have initiated the process as well as retaining the right to cancel it. The only question then is how could a sub-conscious process see into the near future? The process might be analogous to certain quantum experiments where there is a 'delayed choice' in measuring particle positions. When the results are analyzed it seemed the particles anticipated the questions to be put to them. If such a mechanism was possible, evolution would select for it, as it would have great survival value. This 'presentiment' theory may be still rather speculative. It will be therefore interesting to monitor further repetitions of Radin's experiment, to see if it gains in credence.

For planning tomorrow's activities or those in an hour, millisecond offsets are irrelevant. So conscious will power may also exert itself over longer time scales. For example just now I felt an urge to indulge myself by eating cheese and onion crisps, but I 'overcame myself' in willing the urge away. Sometimes I channel such an urge into a desire to eat a nut instead – a healthier alternative, and once one is eating a nut the desire for crisps disappears. Similarly, one often has the choice of continuing an argument or fight, or simply forcing oneself to walk away from the conflict. These are all longer term aspect of will power fed by subtle abstract planning and less by subconscious desires. Note that sub-conscious layers may contain alternatives, but only simple ones, as shown by the cocktail party effect – the main attention may be on the conversation with a friend, and most of the rest is filtered out – but a low level of awareness of other strands of conversation remains so that when one's own name is referred to in another conversation nearby, one suddenly looks up with a start. There are other examples and it is always the case that subconscious thought monitoring can only be of a very simple form like a name: for greater complexity the thought has to enter the spotlight of consciousness: another sign that the ultimate arbiter is conscious thought or will. Thus to escape the subconscious scrutiny of others at a party, keep your phraseology complex!

Another interesting idea that connects free will to quantum processes comes from Stapp (1993), who implies that we could have a superposition of different wave functions corresponding to different choices in our brains, with free will 'collapsing' to one of the alternatives, not through a random process but because there is an element of free will in quantum transitions, even at the level of sub atomic wave function collapses. From outside, an arbitrary free decision would appear random, and on average the quantum mechanical interpretation of a probabilistic processes gives accurate results. Now this theory is a form of pan-psychism, as is Chalmer's idea that every time information is processed, a proto conscious event takes place.

Long-term planning of career, family, holidays etc. involve this long-term free will. Again it is only in consciousness that complex weighing of factors can occur – the sub-conscious is not 'aware' enough to handle this sort of activity, and so these strategic planning processes are fully conscious and more evidence of the efficacy of free will.

The 'God centre' of the Brain, or the 'Brain centre' connection to an 'extramundane reality'?

I wanted to devise a title to this section that would act as a shock to the system of those who have come to expect as a matter of course that anything to do with neuroscience must lead to a 'de-mystification' of everything associated with religion or mysticism. I.e. anything not conducive to a materialistic explanation as fully determined by old fashioned billiard ball atoms must be explained away or ignored. Such is the new religion of scientism. And of course the media obliges: it is automatically assumed, just because a certain location in the brain produces mystic visions when stimulated by electric shocks or epilepsy, that this centre, dubbed the 'God centre', must be a trick that always produces illusions. However, as one neuro-scientist politely explained in a TV interview recently, just because we can produce visual impulses by stimulating the visual centers in the occipital lobe of the brain, doesn't mean that all visual perceptions are illusions! I.e. we have other reasons for supposing that there is a 'real world' out there that causes *most* of our visual perceptions. Brain scans have now also shown that the same visual centers that light up when we see a horse also light up if we imagine a horse in the same position.

So, just as there are various sources that can stimulate our 'terrestrial perception' system, there may be also more than one source that stimulates the 'God centre'. So the assumption that only spurious stimuli can cause a mystic experience is just as incorrect as assuming that only spurious stimuli cause visual imagery. Therefore, the implication is that just as the visual centre evolved to allow us to make our way around a worldly environment of material objects, the 'God centre' almost certainly had an equally important sensory task – i.e. to perceive the immaterial world! Ergo, God, spirits of the dead, angels and other mystic entities may be just as 'real' as the world of matter. Matter is in any case, as we have seen, just a collection of quantum probabilities, the stuff of which dreams are made. So the de-mystifires may have shot themselves in the foot there! I would welcome that, as I abhor their tendency to monopolize the media discussion in the area of philosophy dealing with what Jung termed the 'extra-mundane reality'.

The neuroscientist who voiced the opinion on TV, cited above, is certainly not alone. I was amused and delighted to find evidence for this at a lecture given by a well known German theologian in 2005 in the university town of Darmstadt, southern Germany. The theologian was talking about the impact of findings in neuroscience for theology, religion, and our picture of human nature. He repeatedly referred to ideas as being 'in the opinion of neuroscientists', where he was citing the standard de-mystification party line of the media. Indeed, he sounded not too different to Thomas Metzinger, the well known philosopher, who had spoken in the same series of talks some weeks before that, expounding as usual his idea that free will and the self were all just illusions. After many such references by the theologian, a hand shot up in the audience. A man got up to speak, saying that he had been a neuroscientist for 30 years and in that time did not get the impression that all neuroscientists believed these 'de-mystifying' ideas. He said that every position was represented amongst neuroscientists, from hard-line materialism to dualism to idealism, religion or mysticism, just as in the population as a whole. He pointed out that their researches are at the most basic level and that almost every neuroscientist admits that they are in no position to draw conclusions about these high-level aspects of human culture and human nature.

But of course the media has this image of a scientific priesthood whose religion is exclusively fundamentalist materialism. This impression is given because the books which are publicized are by neuroscientists who conveniently express the expected opinions. Hence our theologian had picked up the impression of a uniformity of opinion where there was none nor could be any. It was amusing to hear him in the rest of his lecture, being extra-careful to replace 'As neuroscientists think…' with 'As Wolf Singer and xxx think…'. Wolf Singer was one of the fundamentalist materialists cited (though he is my no means the most extreme in that regard and may even have mellowed with age, as evidenced by appearances on talk shows and documentaries recently) – I cannot recall the other, xxx, though there are several candidates.

This God Centre of the brain might be a necessary sense akin to seeing and hearing which has atrophied to some extent in the last 200 years as materialism has become ever more predominant. If this is the case, it is a tragedy – it is as if we had willed ourselves blind. It implies that we desperately need to seek out new possibilities for genuine mystical experiences. It could be this instinctive need that drives ever more people to search for a new personal mix in the area of religion and mysticism. This has been particularly noticeable in the last 2 decades as attendance has fallen off more and more in the Christian churches of the west, while the mosques of Islamic countries are filled to overflowing. Hopefully when this period of questing is over, a new equilibrium will be established between mystic and worldly perception and our brain-mind-soul systems will achieve a greater harmony. Such a society might resemble the old Indian civilization where there was a tolerance for experimentation in religious experience.

Note that this idea of a need for genuine enlightenment and mystic experience stands in contrast to the ideas of futurologists such as Karlheinz Steinmueller, who voiced an opinion on this in an interview in the

German magazine Spiegel of December 30th 2006 (see http://www.spiegel.de/spiegel/0,1518,457058,00.html). Steinmueller is of course (since it's a standard magazine that's interviewing him) one of those who consider the stimulus to be the whole story and thus predicts that in some decades we will have pills that selectively stimulate the God Centre to produce a mystic experience ('Enlightenment in the Supermarket' is the title of the article). However, from what was said above, this would be just as meaningless as buying a pill to stimulate the visual centre. If it was a fairly simple visual pill, though more precise than anything we can presently dream of, it would cause arbitrary images in our visual field – probably just like interference on a television screen. Similarly, his cheap and awful mystic pill would just call forth 'mystic noise' of no true value. It is almost certain that only contact with Jung's extramundane reality can produce clear, meaningful sensations in this God Centre.

Just as Homo Sapiens needed creativity to lift him out of the slow lane of hominid evolution into the fast lane we know today, so too he seemed to need a genuine sense of the mystic to stimulate this creative energy.

The fact that all cultures, even if utterly isolated from others, have produced shamans and saints who have had similar visions of mystic beings implies that there really is such an extramundane reality with archetypal beings. If the more mundane evolutionary theory were true, however, and these mystic experiences served to draw the clan together and enforce unity, then one might ask if a simple strengthening of the 'superego' might not have sufficed here – an authoritative figure such as the alpha-male in animal packs. The priests and shamans, on the other hand, were not always of the dominant form that such a superego mechanism would imply. Also, in those early civilizations there was also the rise of kings, who sometimes justified their power by association with the gods. But this was not necessarily the case – normally military might coupled with patriotism was sufficient to ensure loyalty. There was no special evolutionary need for a mystic dimension to the power structures. Rather it is as if, since this mysticism was a vital aspect of early human reality, the mystic was co-opted by rising political powers to emphasize their authority.

So we see that as so often before, the materialist-fundamentals have got the cart before the horse. The media, in its usual rush for a statement from the yes-men, has seized on a poorly argued case. The old days of man bites dog style journalism are over, at least where science and philosophy are concerned, maybe because journalists feel more at home with politics.

Genome of a few Megabytes – the Emperor's new Genome

Finally, turning to evolution and genetics, it is time to examine the implications of the decoding of the human genome which revealed that the numbers of genes was far fewer than expected by earlier predictions. These predictions had been based on estimates of the complexity of the human body and brain. In particular, as neuroscientists in the 1980's and 1990's discovered ever more precise locations for modules governing in amazing detail many of the cognitive functions of humans, the number of genes controlling brain structure was expected to be in the hundreds of thousands. Imagine the shock, then, when it was found that the total number of genes was no more than 25,000. In 2012 the evidence had mounted that it was very little more than 20,000 – maybe 23,000. In the late 1960s, predictions estimated that human cells had as many as 2,000,000 genes, as that was thought necessary to explain brain/body complexity,.

The lower complexity of the genome also has implications for the claims of evolutionary psychology, as espoused by writers such as Dennet and Dawkins.

While the consensus amongst scientists is that Darwin's theory is basically correct, the old school of thought regarding natural selection and the 'selfish gene' is open to re-interpretation. Contrary to what is claimed by Dennett (1995), there remains some latitude in interpreting the details of the theory. This is discussed in the next section. First, let us turn to the genome and examine its structure.

The Structure of the Genome

The DNA chain that makes up the genome is essentially a ladder whose rungs are made of 'base pairs'. The bases, Adenine (A), Guanine (G), Cytosine (C) and Thymine (T) (see the graphic below). Each of these bases is a simple molecule of a specific form. These forms are such that A can only bond with T and C with G. Thus the rungs are either of form A-T, T-A, C-G or G-C. Reading along one side of the ladder, the base sequence can be any combination such as ATTGCAG. The other side of the ladder is then the complement, with each base connected to its partner, e.g. for the latter example TAACGTC. One side of the ladder is called an RNA sequence. When the RNA is paired with its complement, the resulting ladder is the DNA.

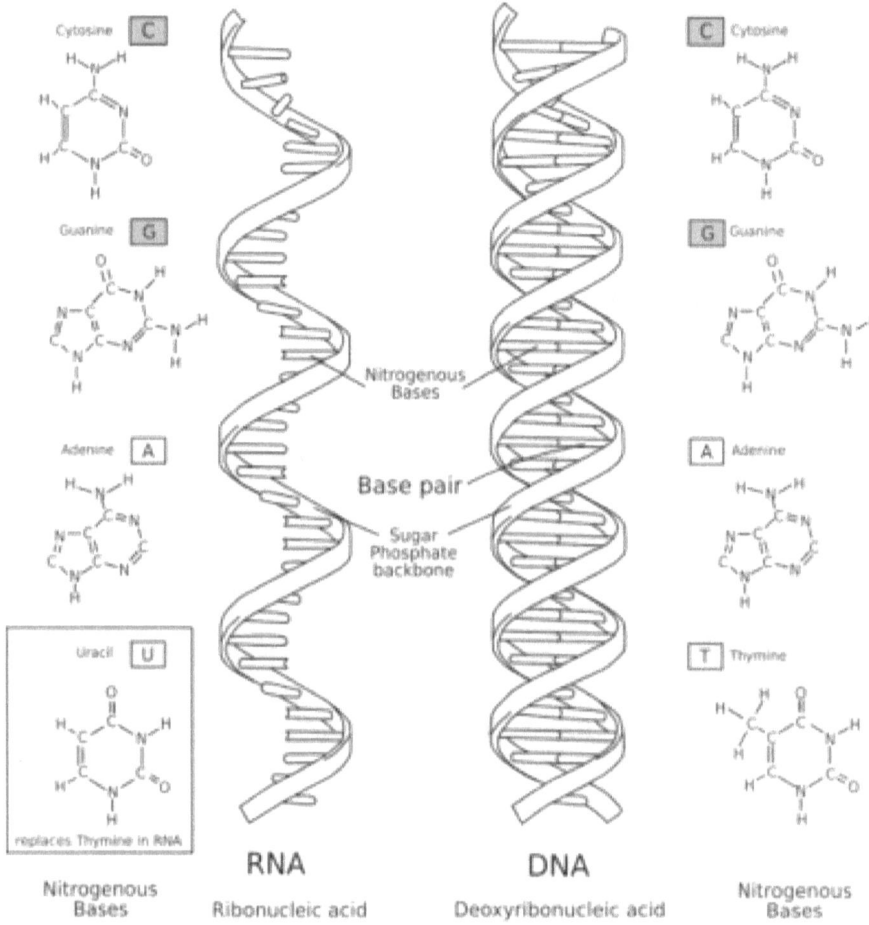

Figure 15: The structure of DNA or RNA in terms of its base pair constituents. (image from Wikipedia).

Now the number of rungs or base pairs in the genome is about 3,000,000,000. This is enormous, and at first it appears as if the genome has an enormous amount of information. However, there is much repetition and many 'nonsense' sequences. Only certain islands of complexity, i.e. the genes and maybe some other sections, actually have hard information. The genes are special because they control formation of the proteins, which are the building blocks of all the different cell types in the body.

The DNA is arranged in bundles called chromosomes. For example, the human genome is made up of 23 pairs of these chromosomes. Each chromosome contains from hundreds to thousands of genes: some of the chromosomes are surprisingly sparse, being mostly 'junk', while others are relatively dense with genes. Men and women both have 22 chromosome pairs in common. It's the 23rd pair that makes the difference: women having 2 copies of the so-called X chromosome while men have just one X and another, smaller one called the Y chromosome. So when you hear people taking of the totally different genomes of men and women, remember that they are essentially identical in 45/46 or 98% of the genome – only 2% is different: it's a case of vive la différence!

Note that this way of determining the sex of a child is not the only way. E.g. the central American fish 'swordtail' has three different types of gender chromosomes, a X, Y and Z! It gets complicated, as males can have YY or XY whereas females can have XX, YZ and XZ. Then parents with XX and YY will always have sons, while parents with XZ and XY will have daughters 75% of the time and sons 25%.

Now if we consider the genes as words, they are further sub-divided into sections of 3 'letters', where each letter is a base (A, G, T or C). These 3-letter sections are called codons, and when the gene gets translated into a protein, these codons are first translated into 'amino acids', which are then the building blocks of the protein that corresponds to the gene. See diagram below:

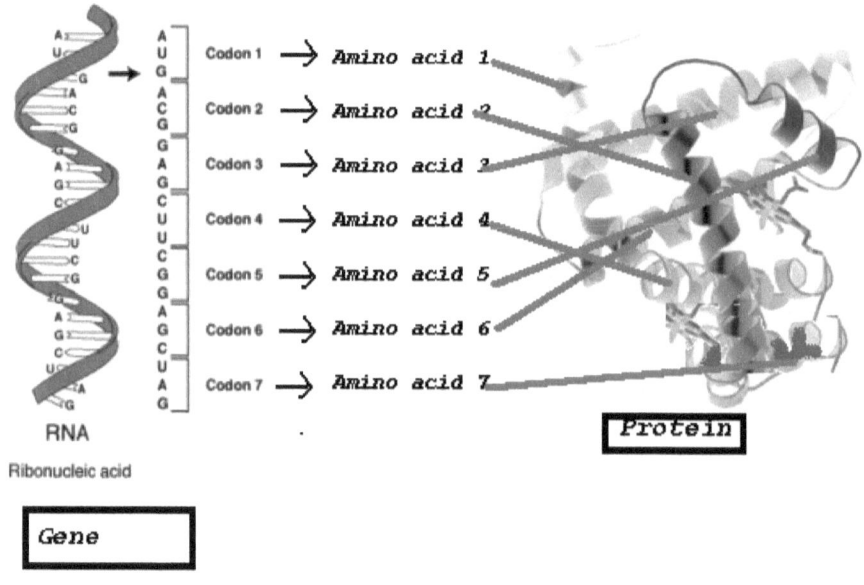

Figure 16: How a gene gets translated into a protein. (image adapted from 2 images from Wikipedia).

From, e.g. http://home.austarnet.com.au/stear/case_against_perl off_ac.htm , we get the following estimate for the information content of the genome:

"Human beings are far more complicated than bacteria, with about 10 to 20 times the number of genes. The human genome is encoded in 3 billion base pairs - or 750 megabytes of ASCII. But human genes are coded in only about 1.5 % of our DNA - some 11.25 megabytes of ASCII."

In a later section we will examine the structure of the genome in some more detail in an effort to estimate its complexity. First, let us look at the old view of the 'selfish gene', still prominent in many books on the subject.

The old version of Darwinism and the 'selfish gene'

Some writers insist on adhering to the old view of the selfish gene first made popular by Dawkins and later taken up by Dennett and other authors. One of the latter is Robert Winston, popular presenter of BBC series such as 'Walking with Cavemen'. In his book 'Human Instinct' (Winston 2002), although making it plain that genes alone cannot explain how humans and other animals behave, he still ascribes great significance to the thesis of the selfish gene. Not only that, but the general tenor is reminiscent of Dennett in the way that he emphasizes the merciless, savage nature of evolution. He prefers to think that war is something that is inevitable due to influences from our genes. However, although he does make a good case for there being aggressive tendencies in humanity caused by certain combinations of genes in the human genome, in my view he fails to make a convincing case for the inevitability of war. He cites studies that show that women with only one X chromosome are more aggressive if that chromosome comes from the mother's genome. In a normal woman with 2 X chromosomes the father's X chromosome has a moderating influence on the aggressive tendencies due to the maternal X. By the same token, men, possessing only one maternal X, are pre-programmed for aggression. However, it is a far cry from this to the idea that war is inevitable. Winston implies, by citing cases of murder and war, that these expressions of violence are the norm. But the actual instance of them is actually rather rare. Most people will go through life without ever knowing war or murder. In Japan, for instance, the rate of serious crime is extremely low. Quoting the wikipedia article on 'crime in Japan':

"Major crimes occur in Japan at a very low rate. In 1989 Japan experienced 1.3 robberies per 100,000 population, compared with 48.6 for West Germany, 65.8 for Great Britain, and 233.0 for the United States; and it experienced 1.1 murder per 100,000 population, compared with 3.9 for West Germany, 9.1 for Britain, and 8.7 for the United States that same year."

These statistics show both that there is in general only about 10 or less in 100,000 chance of being murdered in any given year, but that cultural factors may reduce the level dramatically, in the case of Japan by a factor of 10. Thus to imply that the genes impel people to murder is something of an exaggeration. Winston does succeed in making the case that murders of children by parents, which are extremely rare, are biased towards murders where the children are not genetically related to the parent (i.e. the 'evil step-mother' syndrome). However, the number of re-married parents who never harm their children is not cited. It is obvious from the above statistics, however, that as child murders are rare compared to those of adults, there will be less that one in a million chance of them in any given year. But the number of families where one parent is not the biological parent of some of the children is rather high, since, as Winston cites in ' Human Instinct', up to 40% or 50% of marriages in many countries end in divorce, of which a sizable portion will re-marry, of which again many will have children. Thus there must be of the order of 10% of children in this situation, and yet only one in a million might be murdered. So the emphasis must rather be on the extraordinary strength of the inhibition against such gross violence. In another section of the same book Winston redeems the image of human nature somewhat in citing evidence for altruism. It does seem just as likely that we evolved from gentler, Bonobo-like chimps as from the normal, more aggressive ones. Normal chimpanzees have been observed murdering members of other groups. But Bonobos, primarily fruit eaters, are not as aggressive.

If we look at the evidence that Homo Sapiens lived for tens of thousands of years in harmony with the Neanderthals, both in the Middle East and in Europe, before the latter died out, it seems that at least 30,000 years ago there was no tendency to major warfare or pogroms. It is likely that war is a relatively recent invention. If we are to believe feminist revisionism regarding the ancient history of Europe, there may have been a lengthy period where a Mother Goddess was revered and a form of matriarchy prevailed. Reina Eisler argued, in her book 'The Chalice and the Blade' (Eisler, 1987), that during this era of the matriarchy there was peace for many hundreds if not thousands of years. She supported this by citing the archaeological evidence of Marija Gimbutas. Though some have accused Gimbutas of over-interpreting the data, it would seem as if the main waves of war came with invasions of 'Kurgan' tribes from the east. There were certainly many statues of plump women from before this period, indicating continuity in some sort of cult that revered what seemed to be a goddess or maybe just female fertility.

Figure 17: The pyramid complex of Caral, Peru. It is approximately 5000 years old, making it the oldest city complex in the Americas. (Image from Wikipeda).

Some additional support for the idea that there were intervals when war was almost unknown comes from the archaeological finds at Caral, in the desert of Peru. This remarkable site seems to be the oldest city complex in South America. At 5000 years old, it is much older than any other complex on the American continents, north or south. Ruth Shady (Shady et al., 2001) was the archaeologist who did most to bring it to the attention of the scientific community. Quoting the Wikipedia article on Caral:

"Unlike most cities, no trace of warfare at Caral has been found; no battlements, no weapons, no mutilated bodies. Shady's findings suggest it was a gentle society, built on commerce and pleasure. In one of the pyramids they uncovered 32 flutes made of pelican and condor bones and 37 cornets of deer and llama bones. They also found evidence of drug use and possibly aphrodisiacs."

Thus at least in South America there was a time when a peaceful civilization existed based on trade and agriculture. The same could indeed have been the case in Europe, maybe thousands of years before that. Maybe only after some time, some cities or states grew powerful under rulers whose lust for booty led them to wage the first wars. Maybe this did involve a transition from a matriarchy to a patriarchy, at least in Europe. It would be nice to think that we can return to this primeval state in the near future, if technology allows poverty to be eliminated and the cause for the remaining wars to be removed. Though of course material wealth alone will not keep a nation peaceful, as we have seen from certain rich aggressors unleashing arbitrary wars on poorer countries (e.g. the Iraq war). Coupled with an increase in wealth there has to be a cultural and moral development. One of the problems in human history is that these factors have not been in step – the most dangerous example was the cold war: nuclear weapons were developed that could destroy the world several times over, but the aggressors were still relatively infantile in their emotional development. Luckily, game theory was used by both sides at the right time and a catastrophe was averted.

In 'Human Instinct', Winston, though espousing the theory of the selfish gene, does show an appreciation of something beyond science in his final chapter on 'Morality and Spirituality – beyond instinct?' This would lead one to think that maybe the idea of the gene-rous gene might be more appropriate. Forgive the pun, but after all the genes are a gift of nature, and it is really what we make of them that counts. Human culture in the last few thousand years has started to evolve at a more rapid pace than purely biological evolution.

An indication that Dawkins' view is becoming less popular was when in August 2007, the ex-editor of New Scientist, Nigel Calder, in an interview for the Science Show (http://www.abc.net.au/rn/scienceshow/stories/2007/2001959.htm) attacked his view:

Robyn Williams: Give me a couple of examples of current orthodoxy that you think is somehow errant and will be exposed as such.

Nigel Calder: One easy one is Richard Dawkins. His account of evolution is hopelessly out of date. There are all kinds of things that happened to genes that just don't figure in his way of thinking. There are all kinds of ways in which accelerated evolution can occur involving several genes at one time, and yet the idea of the single mutation being tested by natural selection, which has been the dogma for 70 or 80 years...I mean, it's dead, defunct, but the people who are discovering the other things just don't get reported very widely, even though they are very distinguished scientists themselves. That to me is an example of where a top expert is wrong."

I will come back to the theme of instinct and the genome later, in a discussion of the difficulty of explaining the complexity of instinctual behavior in the light of the 'missing genes' high-lighted by the decoding of the human genome.

There is no conclusive evidence that natural selection alone is responsible for trends in evolution. Other factors, such as subjective consciousness, may also be evolutionary drivers. Llinas (2002), in an essentially conventional analysis by a leading neuroscientist, makes a good case for this, showing how subjective consciousness may be an important survival factor (a zombie may not feel the same impetus to flee from a charging sabre toothed tiger). He even suggests that subjective consciousness may be present quite far down the evolutionary chain. Thus it may be necessary to have subjective consciousness, with its associated emotional drives, to power volition and reactions to danger, hunger, etc. Is it so incredible to think that flies have a sort of subjective awareness of danger? Try looking your cat in the eye and denying her that!

Note that consciousness and intelligence are interrelated, as it's useless being conscious of something but not knowing how to deal with that thing. E.g. If a lorry is approaching very fast on a narrow lane and one will be killed if one does not get out of there, and there is a garden beside you, as a cat you will jump up on the fence and into the garden. That's a very skillful action involving coordination of muscles, sense of balance, planning, timing etc. This is a form of 'animal intelligence'. A human would turn the handle on the gate and enter the garden in that way. A trivial example, but it shows that different solutions to problems are arrived at by different forms of intelligence. I won't discuss intelligence at length here, as the emphasis is on consciousness. But when you consider how difficult it is for a robot to be programmed to open a fridge and take out a bottle of milk and pour some into a glass, you realize that this is also an act of great intelligence. Yet IQ doesn't measure this basic ability, which is linked to how consciousness can learn different patterns of behaviour and make them automatic.

The missing genes

In discussion of the effect of genes on behavior or on development, it is customary among commentators to dip into the genome as if it were a cornucopia of latent possibilities, instructions and information. But as soon as one starts to do a little calculation of the actual information content of the genome one runs into some trouble. Winston (2002) cites evidence that in brain development the level of complexity of the different functional modules and their interconnections precludes a detailed description in the genes. There simply is not enough information present. Let us try to quantify this a bit – as hard figures are a rarity in discussions of the genome, it would seem.

The latest estimates of the number of genes in the human genome is about 22,500 (mean of lower and upper limits of 20,000 and 25,000). What does this mean in terms of information, as we are accustomed to think of it in computer files and programs, as measured in bits, bytes and megabytes? First, let us recap on the structure of the genome.

Let us examine how these figures for information content are arrived at. If we think of the possible choices for one side of each rung, there are obviously 4, as we can choose bases A, T, C or G. So how many bits of information is this? Recall that one bit is the smallest unit of memory in a computer: it can either be 0 (low) or 1 (high). So to label 2 possibilities we need 1 bit, and to label 4 possibilities we need 2 bits. So each rung of the DNA ladder encodes 2 bits of information. We may see this more easily if we look at 2 bits being able to take values 00, 01, 10 or 11, which we can arbitrarily associate with A, T, G or C respectively. The other side of the rung gives no new information as it is fully determined by the first side. Now one byte is simply defined as 8 bits. So to get one byte we need to take 4 rungs of the DNA ladder, to give 4 x 2 = 8 bits = 1 byte. Thus in principle there are 3,000,000,000 / 4 bytes in the genome, or 750,000,000 bytes = 750 megabytes. In fact, there are somewhat more than 3,000,000,000 base pairs in the genome, but the figure of 750 MB is often quoted and is the current estimate used in Wikipedia as of Oct 15th 2008.

The 22,500 genes are embedded in this raw data of 750 MB, but are like currants in a fruitcake and make up just 1.5% of the total cake.

I.e., of the 750 MB, only about 1.5 percent consists of 'Exons' or the protein coding sections of the genes. Separating the Exons are sections of non-coding DNA called 'Introns'. The Introns comprise about 5% of the genome. Human genes have on average between five and 178 introns. They act essentially as place markers between different arms of the gene, so during the process when the protein is finally folding, it will often fold at sites corresponding to the introns – they may thus be thought of as hinges in the origami of the folded gene. Also, for large proteins, they may split along one such 'hinge' and join up with a section of another protein, to give another member of that protein family.

More than half of the genome consists of repetitive sequences, or "junk DNA,". Most of this junk, or about half of the human genome, consists of repeats of short (between 2 and 300 bases) sequences. This repetitive aspect implies that the function of most of the junk is minimal, or non-existent. By comparing the genomes of humans with those of our evolutionary ancestors, it seems that the junk DNA has been wandering around the genome for the last three billion years. This wandering could be another indication of non-functionality, as it seems the genome is fairly insensitive to the precise location of the junk.

As for the Introns, it is widely thought that the non-coding DNA (or functional RNAs produced by them), which make up most of the introns, may perform a function in steering processes in the cell. Taking the protein coding Exons alone, we see that 1.5% of 780 Megabytes is about 12 Megabytes. The Introns function is minimal – a larger role in the 'epi-genome' is played by micro-RNAs, which are short sequence of non-protein coding RNA. Their function is in switching on or off genes or gene sequences and this function increases the complexity of the genome somewhat. However, only about 350 are known, with perhaps the final total between 500 and 1000. This is not an appreciable increase compared to the 22,500 genes. Also, the size of the micros is indeed much smaller than the genes, so the total info content is minimal compared to the protein coding genes. Other RNAs are the tRNA, very small t-shaped RNA sections used to guide the protein forming process at the ribosomes. Their complexity is quite small as this function does not vary much and so they do not contribute a significant amount compared to the genes (see Spork 2009 for a discussion of the non-gene 'epi-genome').

One of the greatest mysteries of the human genome most is that it is 200 times larger than that of baker's yeast but 200 times smaller than that of amoeba. This discrepancy in genome sizes is due to different amounts of junk DNA and poor routine housecleaning of the DNA during evolution. The lack of a consistent relationship between the amount of DNA in the chromosomes of an animal or plant and its complexity is called the C-value paradox.

A strong indication that the junk DNA really is junk was given by the experiments of Nobrega et al. (reported in Nobrega M A, et al. (2004) 'Megabase deletions of gene deserts result in viable mice'. Nature 431: 988-93 – see http://genome.wellcome.ac.uk/doc_WTD020724.html) who deleted about 1% of the mouse genome in junk regions and found that the resulting mice were normal in every respect. The 1% deletion case was more thoroughly investigated, but they also looked at a case where 3% were deleted, also with no effect. The mice even had further children who were also normal. Quoting from the above link:

> ' "In these studies, we were looking particularly for sequences that might not be essential," said Eddy Rubin, Director of the JGI, where the work was conducted. "Nonetheless we were surprised, given the magnitude of the information being deleted

from the genome, by the complete lack of impact noted. From our results, it would seem that some non-coding sequences may indeed have minimal if any function."

A total of 2.3 million letters of DNA code from the 2.7-billion-base-pair mouse genome were deleted. To do this, embryonic cells were genetically engineered to contain the newly compact mouse genome. Mice were subsequently generated from these stem cells. The research team then compared the resulting mice with the abridged genome to mice with the full-length version. A variety of features were analysed, ranging from viability, growth and longevity to numerous other biochemical and molecular features. Despite the researchers' efforts to detect differences in the mice with the abridged genome, none were found.

The negligible impact of removing these sequences suggests that the mammalian genome may not be densely encoded. Similar-sized regions have previously been removed from the mouse genome, invariably resulting in mice that did not survive, because the missing sequences contained important genes and their deletion had severe consequences for the animal.'

This result is consistent with the observation that different species of rat, which are very similar animals in almost every respect, can have totally different genome sizes. For example, the Visacha rat has 16 billion rungs in its DNA ladder, while the Brown Norway rat has only 2.7 billion rungs, and yet they are almost identical organisms.

Some stretches of junk DNA did appear to be indispensable in mammals as they were strongly conserved between species that had diverged on the evolutionary tree millions of years ago. That is, though they had no apparent function and were mostly nonsensical (e.g. ACACACACAC ...), they appeared in the same position e.g. in the mouse genome as in the human or cat genome. However, in the study quoted above, it was also found that even when such conserved junk regions were within the area deleted, there was no effect.

The genes are just islands of structure scattered along the chromosomes that constitute the genome. Are they alone responsible for determining the final structure of the organism? The DNA deletion studies have confirmed that most of the junk in the genome does indeed seem to be junk.

Sometimes the junk does have a function, however. Now, a gene is typically not defined by a continuous stretch of DNA consisting only of '3-letter' codons. Instead, it is made up of several sub-sections of consecutive codons, with each sub-section separated by a short section of junk that acts as a place-marker. When the protein is constructed, these marker locations may indicate 'hinges' or points at which the protein may split and maybe join onto another section of protein. In this way, a single gene may be combined in several different proteins.

As if deletion of junk was not enough, it has also been shown in some studies that when some genes are deleted from the mouse genome there is also no noticeable effect and the mice remain healthy (Mestel 1993). This led researchers to the conclusion that there is redundancy in the genome. Many genes are just backups in case something goes wrong with the main system. Thus although they are very useful, their data is not strictly speaking necessary to define the complexity of the body/brain of the animal. In some organisms, e.g. yeast, it was shown that **up to 30% of all genes were redundant !**

Recently, attention focused on non- repetitive sequences such as non-coding RNA. The difference between this non-coding RNA and the genes is that each gene generates RNA in the 'protein code', i.e. as a chain of codons which, as we saw above, are the 'alphabet' of the proteins. These codon chains are recognized by the 'ribosomes' (the cells protein factories) which then convert them into proteins. The non-coding RNA, on the other hand, does not have this codon-chain structure, and so is not recognized by the ribosomes. These RNA sequences do seem to regulate some aspects of gene activity, and so some of their information content could be added to that in the genes themselves to give a picture of the total information in the genome. It is not clear, however, how much of the non-coding RNA is active. When this RNA is added to the tally of information carrying sections of the genome, it does not increase the total information content by more than a few percent of that in the genes and so does not affect the tally of a few MB.

We saw above that the protein coding genes took up about 12 Megabytes of raw DNA 'code'. Due to redundancy and protein folding, however, the actual 'effective' information content is less – about 8 Megabytes. The non-coding DNA, though roughly of the same size as the protein coding part, is unlikely to be anywhere near it in functional information content. To be conservative, we may say that the total non-junk content is about 2 Megabytes, giving a total of 10 MB.

If this were computer code or ASCII code 10 MB would be a fair amount of information. However, compared to large computer programs for complex systems, it is not so much. Many relatively trivial programs use more data storage. On its own it scarcely seems sufficient to specify the complexity of the body, let alone the brain/mind. In fact, if one were to reconstruct the body using a computer program and data files, one suspects that terabytes might be more appropriate, without ever touching the brain/mind complexity. Thus fields such as Sociobiology or evolutionary psychology, which assume that behavior is to a large extent determined by the genes face a major challenge in the light of these findings. Take for example a new born gazelle - it is able to walk and then run within minutes of being born, a skill needed to flee prowling lions. Yet such skill requires sophisticated software, certainly many megabytes in size, to run on the latest robots from Honda or Sony. So the gazelle must have the equivalent of this sophisticated software, but if its genome is only ten or twenty megabytes, then can it all come from the genes while allowing enough data to form the muscles and other bodily components? Can its neural network learn quickly enough to flee a charging lion? The latter is negated for example by the observations of a horse owner friend: "I can't remember how long it takes a foal to run after it's been born, but I think it stands up immediately after birth while the mother cleans it. It doesn't copy other running animals - in the case of the foal sired by our stallion, it was not just trotting an hour after being born, it was racing around the field with its body at an angle of about 45 degrees as it went round corners. In other words this is all wired in, like young cats' ability to see." The latter again begs the question of where all this hard wired behavior comes from in the presence of so few genes. Are we forced to reconsider mechanisms such as Jung's collective unconscious?

In a fascinating recent discussion of evolutionary convergence (Conway Morris (2003)) some interesting points are made:

* Only 20 of the possible 64 amino acids are used by living organisms as the building blocks of the proteins. Three rungs of the DNA ladder, together termed a 'codon', are needed to specify an amino acid. Each

codon can then only have 1 of 20 possible combinations. They are always left handed, just as are the majority of amino acids found in meteorites, whilst simple laboratory experiments produce equal numbers of right and left handed versions. This, coupled with the difficulty of the formation of DNA, suggests that the first stages of life may occur on comet surfaces or elsewhere in the galaxy, later seeding the Earth for life.

* Certain proteins are constantly reused: Rhodopsins, for example, a key type of protein used in the light receptors in the eye, started out with a different function in bacterial metabolism and later was reused for vision in independently evolved visual systems such as those of mammals, insects or cephalopods (e.g. octopus). Thus although the size of 'protein space' is huge, only certain very narrow regions are actually used and reused, implying that there are only a very limited number of possible 'solutions' to life.

* Many organisms are outwardly almost identical, but possess completely different genomes (e.g. the marsupial wolf and the Eurasian wolf, or the Praying Mantis and its look-alikes). Other organisms are outwardly quite different, but possess almost identical genomes (e.g. humans and chimpanzees). Thus the correspondence between a genotype (one specific instance of a genome) and its phenotype (the animal resulting from that genotype) is not as simple as would be implied by a simple Darwinian selection process.

As mentioned above, the estimate of 12 Megabytes for the raw info in the genes was somewhat generous. In fact, since a codon of 3 base pairs is needed to define one of 20 amino acids, it is similar to the letters of the alphabet. It is always found that *only* these 20 of the 64 possible combinations occur in protein coding DNA. That means that the raw 12 Megabytes is constrained. Recall that 1.5% of the 3,000,000,000 rungs in the DNA ladder are protein coding, i.e. 45,000,000 rungs . Since we need 3 rungs to give a codon, it means that there are 15,000,000 of these 'letters'. But as there are only 20 symbols of the codons, compared to the symbols on a keyboard: 26 letters of the alphabet, 10 number keys, and additional symbols like commas, full stops etc. A set least of at least 40 symbols in needed to specify War and Peace or Newton's Principia. Thus really there are only 7,500,000 text symbols in as we know them in the genome. For a typical paperback of less than 200 pages, 1 million such symbols are needed. So the protein coding DNA is equivalent to the information in about 7 standard paperbacks. Adding the non-coding RNA would boost this to at most one extra volume. But more than 8 or so standard 200 page paperbacks are not in there. Quite a slim collection compared to the old TV documentaries proclaiming that the genome had information equivalent to a row of telephone books stretching over the horizon!

In terms of bytes, a set of 20 can be labeled by 5.5 bits or about 0,69 bytes (as a byte is 8 bits). So the 15,000,000 codons of the genome correspond to about 10.4 Megabytes. Again we saw that most of this is hidden in protein folding and so this reduces to about 7 or 8 MB when the non-coding RNA is added.

Operating System, Computer Language and Code all in the Genome?

Coming back to the size equivalent in megabytes (MB), it seems that even with non-coding DNA added there is at most 8 MB in the genome. Even this figure is generous. For a computer to be able to re-construct the body & brain in all its complexity, with developmental sequence leading from fetus to baby to child to adolescent to adult, how big in terms of code would it have to be? Remember that a program on its own is not enough to produce anything – any computer first needs an operating system. Then it needs a compiler for a computer language and finally the code. Okay, the simplest operating system is typically several KB in size. But they can range up to a few MB, and for body-building a more deluxe version may be needed. Similarly, a computer language as simple as Basic needs a compiler of about a megabyte, but anything fancier runs into several MB. Thus for infrastructure alone we need between 1 and 8 MB – the higher figure would more than 'use up' the entire non-coding DNA. So for the code itself we only have at most about 7 MB left over.

If we are to continue with the computer analogy, a logical question to ask is how we would go about programming a computer to produce a complete, functioning model of the body-brain of a baby which would then develop into an adult through all intermediate stages? Does anything similar to this already exist? It is not unreasonable to think there should be, as we already have models of how to build an airplane such as a Boing 747, which has been said to have as many different component parts as are found in the body. However, a Boing is a rigid body with a very limited repertoire of activities, not to be compared to the human body's complex, flexible behavior, capable of dealing with a huge number of actions and unexpected scenarios. Still, how complex is a Jumbo jet? I have worked in the aerospace industry for some years but am more familiar with satellites. Specifically, I worked with satellite software simulators for a time and so have an idea of their size. A good example to take is the simulator for the European satellite ERS-2, whose instruments discovered much about the environment after 12 years of gathering data. One major triumph of ERS-2 was its discovery that the Antarctic ice cap is growing in the centre, while it melts about the edges. That is why Bangla Desh and the Maldives are not going to disappear under water by 2020, as was once thought based on the melting glaciers alone, since the net rise in seal level will be tiny (see e.g. http://www.cpom.org/research/djw-ptrsa364.pdf, http://www.sciencemag.org/cgi/content/abstract/308/5730/1898 , http://earth.esa.int/ers/tenyears/achievements.html).

ERS-2 is a complex satellite, though by no means the most complex compared to some more recent machines. With its power lines, solar panels, instruments, thermal control, data transfer systems etc. it has a number of subsystems. Thus there is a rough analogy to the human body. Though the circulatory system is actually more complex than the power distribution system of a satellite. And the nervous system and brain (even of a baby) are much more complex than the data transfer and computer systems of such a satellite. Nonetheless, there is a rough analogy. But even for this simplified counterpart of the human body the simulator with its support packages needed at the very least about 50 MB to function properly. Below this size it is hard to imagine modeling a complex system of interacting sub-systems. This is confirmed by considering a well known program like e.g. MS Word, which needs more than 30 MB.

The Complexity of the Body / Brain system

A search of the internet revealed only very limited attempts to model certain aspects of the body-brain's activity, such as motion in walking. There do not seam to be any serious attempts to estimate the actual complexity in the body-brain system – at least none that several different searches were able to find. So let us at least list the different sub-systems within the body:

- Nervous system,
- Skeletal system,
- Muscular,
- Cardiovascular,
- Respiratory,
- Renal,
- Neuro-endocrine function,
- Reproduction,
- Digestion.
- Immune system
- Sensory – sight, hearing, smell, taste & touch

That is at least 11 sub-systems, each of them extremely complex and powerful. If 10 Megabytes are to describe that it leaves only 1 Megabyte for each – less, actually, as they still have to be integrated – a lot of information and knowledge must go into the perfect synergy of all these parallel sub-systems, which work together in such exquisite harmony to allow the body-mind to operate so smoothly.

As if all that were not enough, there is also the cellular level: Membrane transport, cascades of proteins involved in complex processes inside the cell and between cells. And all of this must lead to development of baby to child to adult in an ordered, stable manner. Try programming that up!

The way one divides the body into sub-systems is somewhat arbitrary. A table in the Wikipedia article on human physiology (as of Feb 24th 2007) shows yet another division:

.

System	C

The nervous system consists of the central nervous system (which is the brain and spinal cord) and peripheral nervous system. The brain is the organ of thought, emotion, and sensory processing, and serves many aspects of communication and control of various other systems and functions. The special senses consist of vision, hearing, taste, and smell. The eyes, ears, tongue, and nose gather information about the body's environment.

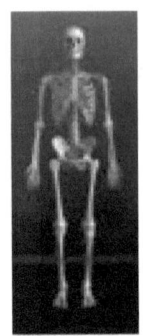
The musculoskeletal system consists of the human skeleton (which includes bones, ligaments, tendons, and cartilage) and attached muscles. It gives the body basic structure and the ability for movement. In addition to their structural role, the larger bones in the body contain bone marrow, the site of production of blood cells. Also, all bones are major storage sites for calcium and phosphate.

osteolo
orthop
disorde

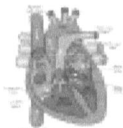
The circulatory system consists of the heart and blood vessels (arteries, veins, capillaries). The heart propels the circulation of the blood, which serves as a "transportation system" to transfer oxygen, fuel, nutrients, waste products, immune cells, and signalling molecules (i.e., hormones) from one part of the body to another. The blood consists of fluid that carries cells in the circulation, including some that move from tissue to blood vessels and back, as well as the spleen and bone marrow.

cardiol
hemato

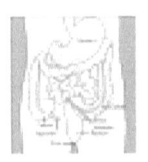
The gastrointestinal system consists of the mouth, esophagus, stomach, gut (small and large intestines), and rectum, as well as the liver, pancreas, gallbladder, and salivary glands. It converts food into small, nutritional, non-toxic molecules for distribution by the circulation to all tissues of the body, and excretes the unused residue.

gastroe

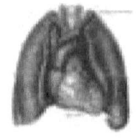
The respiratory system consists of the nose, nasopharynx, trachea, and lungs. It brings oxygen from the air ad excretes carbon dioxide and water back into the air.

pulmo

The urinary system consists of the kidneys, ureters, bladder, and urethra. It removes water from the blood to produce urine, which carries a variety of waste molecules and excess ions and water out of the body.

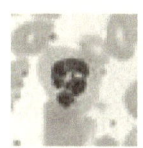

The immune system consists of the white blood cells, the thymus, lymph nodes and lymph channels, which are also part of the lymphatic system. The immune system provides a mechanism for the body to distinguish its own cells and tissues from alien cells and substances and to neutralize or destroy the latter by using specialized proteins such as antibodies, cytokines, and toll-like receptors, among many others.

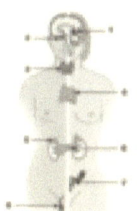

The endocrine system consists of the principal endocrine glands: the pituitary, thyroid, adrenals, pancreas, parathyroids, and gonads, but nearly all organs and tissues produce specific endocrine hormones as well. The endocrine hormones serve as signals from one body system to another regarding an enormous array of conditions, and resulting in variety of changes of function.

The reproductive system consists of the gonads and the internal and external sex organs. The reproductive system produces gametes in each sex, a mechanism for their combination, and a nurturing environment for the first 9 months of development of the offspring.

The integumentary system consists of the covering of the body (the skin), including hair and nails as well as other functionally important structures such as the sweat glands and sebaceous glands. The skin provides containment, structure, and protection for other organs, but it also serves as a major sensory interface with the outside world.

Fig. 17: Sub-systems of the body (from Wikipedia).

Remember that all the systems listed above are reflected in the brain, which monitors and controls them all on either a sub-conscious or conscious level. Thus within the brain there is enormous complexity even at birth, before education has begun. Hence the surprise and amazement of neuroscientists when only 22,000 genes were found – they had estimated that at least 100,000 genes were needed for the brain alone! And although even more complexity of the brain develops only slowly as a child grows, much is pre-ordained or 'programmed', such as the language ability and the location of modules dealing with vision, maths, hearing, reasoning, as well as all those for the sub-systems listed above. Within vision alone there are 40 modules dealing with primary processing such as shape, orientations, speed etc. All of this structure is somehow pre-determined, though not rigidly, as the ability of humans and animals to re-program parts of their brain to take over the task of an injured region shows. However, the re-programmed area co-opted for such a repair job never functions quite as well as the original, implying that what is inherited is a strong tendency for the locations of these modules. Intuitively we would think that 10 megabytes could never cover all of this (basic to survival) complexity. Can we quantify this feeling with some figures?

It is surprising that this sort of quantitative analysis seems not yet to have been attempted. Why is this so? Probably because it is so daunting. But a ball park estimate is not so difficult, as I will now show. This begs the question again as to why it has never been explored before. One factor is certainly the fact that too much information would contradict the simple materialist assumption that all structure is determined by the genome.

We saw in the previous section that a complex man-made system such as a satellite needs typically a minimum of 50 MB to simulate the electronics, instruments, thermal control, data transfer and computer sub-systems. Yet such a satellite is not as complex as the human body. Also, this software simulator does not simulate the 'skeleton' of the satellite or determine how it might be built up from nano-machines, in analogy to the way human bodies are built up from the nano-machines of the cell. Imagine that – a nano-machine that reproduced itself and differentiated in function until it became a satellite. I can assure you that such a system would need thousands if not millions of megabytes, if it is indeed at all possible in the next millennium. Yet this is what the fertilised human egg does in becoming a human being. Thus even if all the non-gene-producing parts of the genome were included to give the full 20 MB of meaningful information, there is no way we can conceive of at present to reproduce the feat of the growth, development and functioning of the human body-mind with programs of this size. With 12 to 15 MB the problem becomes even more acute.

Another indication of the complexity of the body is the fact that there are 210 different types of cell. See the Wikipedia page on this http://en.wikipedia.org/wiki/List_of_distinct_cell_types_in_the_adult_human_body for a reasonably full description. Here, I will list just the 20 main types from the above link:

- ☐ 1 Keratinizing epithelial cells
- ☐ 2 Wet stratified barrier epithelial cells
- ☐ 3 Exocrine secretory epithelial cells
- ☐ 4 Hormone secreting cells
- ☐ 5 Metabolism and storage cells
- ☐ 6 Barrier function cells (Lung, Gut, Exocrine Glands and Urogenital Tract)
  - 6.1 Kidney
  - 6.2 Other
- ☐ 7 Epithelial cells lining closed internal body cavities
- ☐ 8 Ciliated cells with propulsive function
- ☐ 9 Extracellular matrix secretion cells
- ☐ 10 Contractile cells
- ☐ 11 Blood and immune system cells

- ☐ 12 Sensory transducer cells
- ☐ 13 Autonomic neuron cells
- ☐ 14 Sense organ and peripheral neuron supporting cells
- ☐ 15 Central nervous system neurons and glial cells
- ☐ 16 Lens cells
- ☐ 17 Pigment cells
- ☐ 18 Germ cells
- ☐ 19 Nurse cells
- ☐ 20 Interstitial cells

For an example of the sub-divisions of one of these main types, take nr. 9 above, Extracellular matrix secretion cells:

Example of complex detail: Extracellular matrix secretion cells

- Ameloblast epithelial cell (tooth enamel secretion)

- Planum semilunatum epithelial cell of vestibular apparatus of ear (proteoglycan secretion)

- Organ of Corti interdental epithelial cell (secreting tectorial membrane covering hair cells)

- Loose connective tissue fibroblasts

- Corneal fibroblasts

- Tendon fibroblasts

- Bone marrow reticular tissue fibroblasts

- Other nonepithelial fibroblasts

- Pericyte

- Nucleus pulposus cell of intervertebral disc

- Cementoblast/cementocyte (tooth root bonelike cementum secretion)

- Odontoblast/odontocyte (tooth dentin secretion)

- Hyaline cartilage chondrocyte

- Fibrocartilage chondrocyte

- Elastic cartilage chondrocyte

- Osteoblast/osteocyte

- Osteoprogenitor cell (stem cell of osteoblasts)

- Hyalocyte of vitreous body of eye

- Stellate cell of perilymphatic space of ear

If we consider that a cell needs a basic set of about 3000 genes to survive, then the other 19,000 of the 22,000 genes of the genome can be thought of as being distributed amongst these different cell types. This makes on average about 90 (= 19,000 / 210) extra genes per cell type. Note that this extra 90 has to tell the cell how to fulfil its special task and how to integrate this into the organ(s) of which it is a part. We can assume that the non-coding RNA helps with the ordering of the different cell types in the whole. However, as Lazlo points out, the body is in a state of exquisite balance, with synergies of many different kinds operating to integrate the parts into the whole. Does this point to additional fields such as his Akashic or Sheldrake's Morphic fields?

Fractal compression in the Genome, or algorithmic complexity?

As mentioned in the first chapter, some have maintained that the small amount of true information in the genome may be augmented by a sort of fractal compression. As we saw, for some special patterns a high degree of compression may be achieved using fractals. For example, a fern may be modeled by very simple equations. However, if we look at the problem of data compression for images, we find that the early promise of fractal compression proved an illusion, as any departure from the relatively simple fractal structure led to difficulties, so that a standard data compression format such as JPG was as good for almost all images, as long as a reasonable quality was required. Similarly, in the human body, though some aspects may seem to be fractal – e.g. arms and legs protrude from the body in a similar way to the fingers protruding from the hand or foot – in general this analogy breaks down and the structure becomes as difficult for fractals to reproduce as fractal image compression in going from a fern to a complex flower like a lily.

So there is no shortcut via fractals. In the end, the more general concept of algorithmic complexity must be used. I.e. how big must the most compact program be in order to produce the body-brain-mind? How can we generate the body from first principles? The experience with aircraft and satellite simulators shows we need at least 50 Megabytes even for rather rudimentary simulators. To include nano-machine self-assembly of all the complex organs and physiological sub-systems together with complicated synergies would need a lot more. So if anyone ever feels up to the task of a complete simulation of the human body then it will be amazing if they do it using anything less than a thousand megabytes. Even that is a modest figure. This then begs the question – how does the observed complexity arise from a genome of between 10 and 20 Megabtes?

Obviously, a lot more work has to be done to show just how big a program must be to simulate a life from egg to mature adult. Hopefully some time soon, then, a more precise figure may be placed on the 'missing complexity' in the genome. But there is already enough evidence to indicate that this figure will be substantial. But if not in the genome where could the information reside? Some look to epi-genetic material such as the arrangement of the chromosomes and proteins floating the protoplasm of the egg. But experience with cloning shows that this cannot be a very significant factor. It might therefore be necessary to postulate some field accompanying the genome acting in a guiding and steering capacity. This would be analogous to the idea of a 'pilot wave' in some theories of quantum mechanics. Indeed such theories have been proposed – including Ervin Lazlo's 'Akashic Field' (Lazlo 2004) or Rupert Sheldrake's 'morphic fields' (Sheldrake, 1981), discussed in the next chapter.

To summarise: there seem to be far too few genes in the human genome to explain the incredible complexity of the body's diverse physiological systems as well as the brain and its associated mind. Approximate estimates of the raw information in the genome, including the genes and other DNA that does not produce proteins, is that there are fewer than 20 Megabytes, by analogy with computer storage. This is not just some abstract comparison, but a rational assessment of the complexity of the genome. When we try to match this information content with the complexity of the body/ brain/ mind (BBM), we encounter a paradox. The BBM appears to have a complexity far exceeding any machine yet built, and yet we know that the description of complex machines needs far more than 20 Megabytes. The world is still waiting for the first full scale simulation of the body and all its organs, nervous system, brain, circulatory, lymph, digestive etc. systems. If this massive task is ever accomplished, it is certain to need, thousands, or millions of Megabytes. So we conclude that the genome alone is insufficient to explain the complexity of life as we know it. Some other ingredient, such as information fields, 'morphic' fields or the ancient idea of a soul seems to be needed.

Socio-biology, Evolutionary Psychology or a new form of 'Vitalism'?

In the light of the paucity of information in the genome, the agenda of such disciples as socio-biology and evolutionary psychology are on rather shaky ground. Considering the need of such disciplines for a detailed pre-programming of the brain along evolutionary instinctual grounds, the absence of the hundreds of genes thought necessary for the completion of such a pre-programming makes it difficult to sustain the thesis of genetic determinism. Is it not time to question whether a supplement to genetic programming is needed?

The more biologists used reductionist techniques to elucidate the biochemical processes that maintain all living cells, the more they came face to face with the problem of complexity, emergent phenomena and ultimately the puzzle as to how the complexity emerges from simplicity. Despite the desire to reduce biology to chemistry, it is soon apparent that one may no more do so than reduce chemistry to physics. The quantum processes in even the simplest of molecules become rapidly too difficult to solve using the equations of quantum mechanics. Soon, approximations must be used until it is apparent that one is at a new level of order. I.e. chemistry emerges from physics just as biology emerges from chemistry. In both cases the emergence is too complex to be followed by strictly deterministic equations. The emergence in both cases involves non-linear processes, feedback and cyclic processes. In a very real sense chemical processes are independent of the underlying physics and biological systems are independent of the underlying chemical processes.

The further we proceed up the ladder, again and again we experience this emergence of seemingly independent complexity. Also at the level of the mental we may say that mental or conscious processes are largely independent of the underlying biological processes. Except that the mental is so high up the ladder of complexity that there is a sort of double emergence effect, in that the subjective aspect of consciousness accompanies the objective complexity of intelligence, instinct and perceptual processes. As we have seen, subjective consciousness essentially creates separate inner dimensions in a form of virtual reality. In terms of the objective, this is a truly magical process. No wonder that strict materialists are so frightened by the idea of the subjective and tried to ignore and deny it for much of the 20th century. With good cause, since the hard problem of consciousness is unlikely to find a solution in terms of science. Something more akin to magic would seem necessary, since subjective consciousness is nothing less that the soul or atman of the ancient sages.

When Rupert Sheldrake produced the first of his books on his theory of Morphic Resonance or Morphological Fields in 1981, "A New Science of Life", it struck a chord amongst many scientists frustrated with the ultra-mechanistic trends of genetic reductionism. Reductionists were of course outraged at what they saw as a resurgence of the idea of Élan Vital, the old idea that a mysterious non-material essence was needed to explain the vibrant nature of life. The adherents of such a theory were dubbed 'vitalists'. The theory took a few knocks during the 20th century as it was shown that many of the processes in living organisms were based on chemical principles not so different from processes that could be reproduced in the lab. However, the main core of the vitalist idea was the controlling nature of the inner awareness over the outward, objective processes. This conundrum still remains. As the title of Colin McGinn's book says, consciousness in the material world is like a 'mysterious flame'. That's a good analogy, as without subjective conversion of light and sound into the kaleidoscopic experience of conscious vision and hearing, the light waves and air waves would be just blind processes groping their way mindlessly through eternal night, in a despairing pointlessness such as that described by Steven Weinberg at the end of his book about the big bang, 'The first Three Minutes' (Weinberg, 1993).

The emergence in going from physics to chemistry etc. is of one kind, i.e. always going to a new level of *objective* complexity. The emergence of the mental from the physical is of another kind, i.e. the subjective emerges from the objective. It may not be true that this is strictly emergence in that the essence of consciousness might include elements independent of the physical substrate of the brain/body. Also, the emergence, such as it is, is not a one way linear process, as the binding nature of consciousness means that it integrates the results of multiple objective brain processes and at the same time gives them a subjective stamp. We saw this in Chapter 4, in the discussion of the binding problem in neuroscience or neuro-philosophy. As discussed there, a visual image is first dissected and processed in different parts of the brain and then later re-integrated on the 'internal TV screen' of subjective vision, and similarly with the other senses. We then encountered the problem of who was watching this internal TV screen. Old diehard adherents of the behaviorist and allied schools of philosophy, such as Daniel Dennett, are in denial of this process and try to waffle it away as an 'illusion' or irrelevant by-product of objective processes. Others, such as Rudolfo Llinas, recognize the crucial role played by subjective awareness in evolution and posit that it may be present right down to single cell level.

Regardless of how far back along the evolutionary chain we can go before encountering true zombies or automatons, it seems clear at least for us humans that we are not mindless zombies. Thus the idea of dualism is alive and well, and cannot be dismissed as readily as the old school of behaviorists and the, as usual, outdated school of journalists would like to think. This is, in other words, the proof that we are not soulless automata. So if we are not soulless we must have some sort of soul. But the soul is essentially a synonym for subjective consciousness. The only debate then is as to the immortality or not of the soul. Whilst many thinkers are sure that the soul is indeed mortal, there is some evidence from incarnation stories from India and indeed all over the world, that the essential essence of a life may proceed to a new incarnation or beyond. The idea of such a bodily-independent soul could explain some of the missing data in the genome – this might be somehow stored in the 'pilot wave' of the soul. This is a sort of dualist theory of inheritance, which is consistent with a dualist theory of the mind-brain system.

This is the most speculative chapter yet, so while we are speculating why not investigate the implications of this dualist DNA-soul and brain-mind system. It is not such an outrageous proposal as it might at first seem to homo modernis materialis. What are Laszlo's Akashic (Laszlo, 2004) and Sheldrake's morphic fields but synonyms for something like the more traditional (in the west) idea of the soul? A dualist position, where the soul is the driver of the car that is the body, makes sense from an evolutionary point of view, as we have already seen: a zombie is less likely to feel the same sense of urgency in springing out of the way of a saber-toothed tiger as a body controlled by a subjective consciousness or soul. And we have also seen that try as they can, even the most materialist of philosophers will also let the 'ghost of agency' creep back into the picture. The ghost in the machine, so feared by Pinker, Dennett & co. is indeed a ghost or soul. The infinite regress of perception – the question of who is watching the internal TV screen projected in the brain, has, as we pointed out already, no real alternative but the equally dreaded homunculus. And one homunculus or soul is really more economic and reasonable than an infinite regression or Russian doll ad infinitum. The supposed third option of dismissing the impression of subjective consciousness as an illusion is really no option at all, as even an illusion requires a dupe to be deluded – i.e. an audience: which brings one back to infinite regress or homunculus – infinite Russian doll or soul.

The Emperor's new Meme

The idea of memes [Dawkins 1989, Dennett 1995] arose in the heady days before the complete human genome sequencing, when it seemed that genes and natural selection were an all encompassing combination of causes. But memes are on even shakier ground than genes. The short comings of this ultra-mechanist attempt to strike at the core of 'mysterian' consciousness were pointed out in counterarguments in Scientific American ('Meme Theory Oversimplifies How Culture Changes,' Boyd and Richardson, p. 58 and 'People Do More Than Imitate,' Plotkin, p.60, October 2000). Indeed, as the above arguments on the nature of subjective consciousness and genes have shown, the question is still open as to whether the brain (objective) and mind (subjective) may be equated. Thus rather than considering such questionable concepts as Memes and similar ultra mechanist/materialist ideas, it might be appropriate to focus on the mysteries of subjectivity and top down, non linear approaches. Otherwise the 21st century may close with a book titled 'The Emperor's new Meme' just as the 20th closed with 'The Emperor's new Mind'.

In fact, memes are just an attempt to divert attention from such aspects of consciousness as meaning. Meaning is a far more fundamental and mysterious quality – no program has been able to convey this aspect of abstract thought. Indeed, abstract thought, which separates us to a large degree from other members of the animal kingdom, is another mysterious aspect of human consciousness. And memes are nothing more than another name for ideas – and ideas are dangerous, which is why the zealots of ultra-materialism must contain them by labeling them with something that lets them appear more mechanistic than they are. But in fact ideas are ethereal. William James, almost 120 years ago, gave a famous description of some of them as like birds flitting from branch to branch. The more solid of the ideas corresponded to the branches. As Deepak Chopra says in his book 'Life after Death', there is no clear notion of how we flit from idea to idea, hauling up exactly the right memory for each occasion. Or in creating each new paragraph, we are assembling ideas never before expressed in exactly this manner – but the right words come one after the other: some of them hackneyed, others not. The hackneyed ideas might be those that the materialists jump on with glee, labeling them as 'memes'.

Considering the evidence presented in previous chapters of the information shortage in the genome, the idea that genes are all-important is shown to have taken a serious knock. Thus this is no time to up the ante for the case of mechanistic determinants by introducing dubious and superfluous concepts such as memes, where the old idea of ideas does just as well. But idea is too free, and has associations of idealism, in the philosophical as well as inspirational senses. Thus the mechanistic camp seeks to replace the outdated idea with its stillborn concept of the meme. It would be better to concentrate on an exploration of the mystery and wonder of the age-old concepts of the idea, of the abstract, perceptual binding, meaning, qualia etc. I.e. all the mysterious contents of subjective consciousness.

More things in Heaven and Earth

A cause celebre that would clinch the argument against reductionist materialism would be the reality of psychic phenomena, for despite Blackmore's (2001) claim that machine consciousness could also produce psi (psychic ability – i.e. either paranormal cognition (ESP (Extrasensory perception), remote viewing, etc.) or paranormal action (psychokinesis, etc.)), it would, if it exists, seem to be a property of conscious (or sub-conscious in the case of poltergeists or prophetic dreams) will-power. As in any paradigm change, the anomalies must mount up to the point where the old paradigm is no longer capable of explaining many observed phenomena.

Note also that proof of the reality of psychokinetic powers would be a major nail in the coffin of the materialist hypothesis for another reason. Take, for example, the argument of Baggini (2005) that the world is causally closed within the material paradigm. He therefore justifies skimming over such mechanisms as quantum processes for rescuing free will as he sees in them merely a way for random processes to influence the predetermined course of matter and forces. However, if the mind can be shown to influence dice at a distance or shatter glasses, light bulbs and other crockery as in poltergeist phenomena, then this is evidence for 'property dualism' – i.e. the coexistence of matter and mind mutually influencing one another but made of separate 'stuff '.

From there it is a logical step to imagine mind influencing the neural processes in the brain to control the body. This should be an easier matter than influencing dice or other external matter, as it is a case of 'internal psyhokinesis', if we imagine the mind as growing used to exclusive association with one brain as the individual grows up. We could even, in this case, imagine mind influencing quantum probabilities to steer the firing of synapses. The latter should be even easier than bulk psychokinesis as it does not violate conservation of energy and thus the quantum processes would be a lever for easing the ability of mind-matter interaction. However, recently even the conservation of energy (COE) constraint has been challenged. By the time this book is published, the magnetic power system of STEORN ltd. of Dublin, Ireland should have made world news: As one of a handful of scientists in the original forum of the company who were testing the technology, I can speak for the rest in that 2008 started with exciting demonstrations to us at the company site in Dublin. Thus the sacrilege seems true – COE really is broken, just as conservation of parity was found to be broken for some sub-atomic processes. Thus mind may indeed interact even more robustly than the quantum probability way with brain processes.

In Baggini's book there is also a chapter with the same title as the current one, but its tone is decidedly different. It indicates that many people put their faith in God or some other extra-mundane reality, but they are sadly mistaken in thinking that there is anything other than the material world of the normal senses. As for the sixth sense, he dismisses this out of hand as lacking in any sort of credible evidence. These two areas go hand in hand to a great extent, for though paradoxically it was the church which persecuted any sign of psychic powers as being satanic, in fact proof of psychic phenomena would be the chink in the armor of materialism after which the next step could be to infer a spirit or extramundane reality. So in fact Baggini fails to consider adequately that which may lay outside his philosophy as his attitude to anything in Heaven or Earth of that sort is to ignore and dismiss it as unworthy of consideration.

There is also an annoying habit of evolutionary psychologists or neuro-materialistic authors such as Steven Pinker to boast in their introductions that they will, in their review of neuroscience and evolutionary biology, explain why we choose our mate and why some people are disposed to believe in the paranormal. Such authors are keen to point out that there are regions in the brain associated with belief in paranormal or religious phenomena as well as with the actual seeing of supernatural entities. However, they neglect to mention the converse: a deficient amount of tissue in the relevant brain regions may equally well be construed as a deficit in the ability to believe in the possibility of the paranormal or to see ghosts etc. Since all brain modules are presumed to be the product of an evolutionary process, it seems that we need the modules for seeing, hearing, language, hunger, sleep, face recognition and reflexes such as avoiding fire. In the same way, more subtle modules such as the instinctive physical understanding of how a ball flies in a parabola are useful. It also seems that evolution found it useful to have regions of the brain conducive to religious feelings and mystic visions. In the same way it appears to have been necessary to have a module for belief in paranormal abilities. Maybe the reason for that is that there really are such phenomena, though the materialistic school of neuroscience or evolutionary psychology would deny this. But such a denial is inconsistent considering that brain tissue is not normally devoted to something that is useless and counter-productive from an evolutionary point of view. The reasoning outlined here, though, is not part of the things in Heaven and Earth which they are prepared to consider in their philosophy.

Magazine articles quoting these sceptical views are just another example of a hidden agenda being smuggled on-board the modern consensus paradigm in the form of defining arbitrarily what is acceptable in Heaven and Earth or to society. Must 'normal' inevitably be equated with straight-laced ultra-materialist, completely lacking in the 'paranormal module' as in the simplistic popular view of the stereotyped scientist? Is the freedom to choose one or other philosophical stance to be denied us by the new arbiters of orthodoxy, the modern puritans such as Dennet, Dawkins, Damasio & co.? Which sci-fi writer was it who wrote of 'creeping conformism'? The phrase describes well the increased tendency to make life extremely difficult for anyone who steps outside the straight jacket of scientific orthodoxy.

The greatest sinners against the new orthodoxy are the parapsychologists. Yet it is surprising how many scientists were secretly fascinated by reports of psychic phenomena. William James, the 'father of modern psychology', had a life-long interest in the paranormal. For many years he was president of the American Society for Psychical Research and he personally thoroughly investigated the famous medium Leonora Piper, concluding that she was the 'white crow' that he had sought for many years. It is almost certain that you won't have heard this story, as the skeptics who flood the modern media with their material are peculiarly selective: with them there is a case of permanent short term memory only. Any anomalous material older than a few years is ignored. Thus much that was remarkable about the last century and a half is forgotten and a sanitized picture of orthodoxy is presented to the world. But history was not that neat, and there were many giants of science such as William James who made a serious study of the paranormal and concluded that there were things in Heaven and Earth that were not dreamt of in materialist philosophy. Freud was another who was fascinated by psychic abilities, as was his rebellious pupil Carl G. Jung. Jung openly wrote of his interest in psychic phenomena, for example in his book 'Synchronicity', which coined that very term. Freud, however, fearing censure of his academic peers, kept his interest private.

There were indeed superstars of the psychic scene in the 19th century, the Uri Gellers of their day. There was, for example, Leonora Piper, a trance medium who as mentioned above impressed William James. Then there was the true analogy to Geller, Daniel Dunglass Home, who was impressive for several reasons. First, he performed most of his séances in daylight, not in a darkened room as did most mediums. Secondly, he could perform in the houses of his hosts without any chance to prepare the venue as he could on his own territory. Third, his abilities were not just information as in the case of Piper, but involved psychokinesis such as lifting massive tables or causing an accordion to play in a cage under the table. Some famous physicists were impressed by him. For example, Wiliam Crookes, inventor of the Crookes tube (on which the first TVs (before the recent plasma and LCD technology hit the market) were based with their 'cathode ray tubes') and discoverer of Thalium, witnessed some of Home's feats of levitation, where he appeared to float out a window of Crookes' house or lift up to several feet off the ground. See the Wikipedia article on Home for a balanced account. As mentioned there, though suspected of fraud on some occasions, this was never proved. Of course, it would have been better if he could have been examined under modern laboratory conditions. But the fact that he convinced leading scientists, politicians and royalty of his abilities by daylight and was feted at the courts of Europe for many years speaks for the possibility of his abilities having been to some extent genuine.

Uri Geller himself remains controversial. He has been caught cheating on some occasions but claims he needed to supplement his genuine talent to maintain audience interest. The problem is that he is a showman and thus not the ideal scientific subject. Nevertheless, on some occasions he did perform well on TV without any sign of trickery. If even one such 'white crow' performance could be verified, this would suffice. He impressed the physicists Targ and Puthoff, who were partly funded by the CIA in their psychic studies at the Stanford Research Institute, one controlled study of which was the basis of an article in Nature (Targ & Puthoff 1974). Note, though, that the evidence presented in the Nature article was for Geller's telepathic reproduction of drawings, and not his metal bending, which he did do at SRI but not under conditions well enough controlled for the physicists to include it in their report.

To a great extent, the effect which Geller had on TV audiences was more impressive than anything else about him. Thus following his appearances on TV all over Europe in the 1970s, a group of viewers, mainly children or adolescents, claimed that they too could bend metal. Some of these children underwent strictly controlled laboratory tests and were occasionally successful. For example, The Munich physicist H.D. Betz describes a test with a 12 year old girl from the Bavarian countryside (translated from a section quoted in Von Lucadou, 1995):

"As test objects, metal strips of aluminum, iron and copper were prepared. Typical measurements were 120 x 12 x 1 mm. All the strips were marked with engraved numbers. As the girl to be tested, S., was first confronted, without prior warning, with these standardized objects, the following sequence of events took place:
The author (Dr. Betz) laid one of the test strips in the

middle of the table. The surface of the table consisted of a thick, level slab of glass and the strip lay totally flat and level upon this surface. S. now held down one end of the strip, A, with her left thumb by applying downward pressure. She then began to gently stroke the region of the strip near A, i.e. region B, from above with her right thumb. As soon as S. touched the strip with her right thumb, the other (free) end of the strip, C, began to lift up from the table surface, due to bending of the strip at B. The end C rose with nearly constant speed of an estimated 0.5 to 1.0 mm per second. After less than a minute C was 60 mm above the table surface - i.e. a bending angle of 50 degrees was achieved."

The usual trick of switching a prepared strip of one's own was ruled out by the marking of the metal. This country girl was not a professional magician or performer like Randi or Geller, and so other sophisticated tricks can be ruled out. She also couldn't apply a special gimmick as she had no idea before the session what was to be asked of her. The trick of distracting attention while one 'pre-bent' the strip was also out of the question as the degree of bending was clearly visible at all times.

Such tests with the mini-Gellers are indeed more convincing than with Geller himself. I also discovered to my surprise that one of my neighbours has a spoon that he says he bent a la Geller during one of the TV shows in the 1970s. I am inclined to lend him credence as he always struck me as an archetypically conservative citizen.

Also more convincing than Geller himself are perhaps poltergeist phenomena, which appear spontaneously and are usually associated with troubled adolescents. Some cases are impressive. For example, that which occurred in Rosenheim, Bavaria, in 1967. This was impressive as surges in electrical supply were thoroughly investigated by the town authorities, who went as far as disconnecting the lawyer Sigmund Adam's office from the net and hooking it up to a dedicated power supply in an adjacent courtyard. Despite this, the power surges continued, and physicists from a Max Planck Institute centre measured anomalies on their instruments. Also, witnesses including policemen saw pictures rotating in their frames, lamps swaying and then exploding, and a massive cupboard move out from a wall by up to 30 cm, though no one was touching it. Though nobody seemed to be phoning from the office, the phone bill was huge: the records showed more calls to the time service per minute than could have been made by dialing with the analog technology of the 1960s. Hundreds of such calls were registered, though the phone system was locked by Adam. And all these anomalies only occurred when the young office employee Anne-Marie Schneider was present. Brtitish TV made a documentary on the case in the 1970s and recently the main official German channel ran a program on it – see e.g. http://www.daserste.de/dimensionpsi/gaensehaut_08.asp . It would seem essentially impossible that Schneider could have caused all these phenomena by normal means. Also, similar phenomena occurred in other places where she was present.

Another more recent case that is similar in its involvement of police, officials and scientists in the investigation was that of the mysterious fires in Canneto di Caronia in Sicily from 2004 - ? (uncertain if they have ceased for good). In this case, household appliances would burst into flames, some of them even when disconnected from the mains supply and after the village power had been cut. Cars would lock themselves. Scientists investigating the case dismissed mechanisms such as heat or electric forces due to Mount Etna. No such mechanism seemed capable of explaining all phenomena such as cars locking and all the fires.

There is a certain consistency in reports of poltergeist phenomena since antiquity. Already in ancient Egypt flying stones and shaking furniture were recorded. Often stones are reported to seemingly come from a wall, float slowly across a room and drop to the ground, sometimes just before they might have struck a person, as if avoiding injury. Usually such stones, when lifted up, are found to be very warm. Furniture is often reported to move as if by a ghostly hand, plates and other crockery shatter unaccountably and knocks and bangs are heard. It is the consistency of such aspects of these cases across centuries and cultures that implies that there might indeed be a core of truly psychic phenomena behind them.

Note that I see the greatest possible evidence for psi in naturally occurring or spontaneous cases. For this is the real nature of psi. To try to tame it and bring it into the lab has been shown to be the wrong way to go. The history of the last 60 or more years shows this. The trend toward attempts to set up reproducible laboratory results started with J. B. Rhine in the 1930s. It was Rhine who coined the term ESP (extra-sensory perception) and started the scientific discipline 'parapsychology'. His tests with card guessing and dice throwing trials gave results that were well above chance. However, sceptics dreamt up ever stricter security measures to rule out cheating. Some of the improvements in the experimental setup were justified but others doubtful. The end effect was that the stricter the rules and consequently the more 'boring' the trials, the smaller was the effect, until it was barely above chance.

Then from the end 1960s onward, a new even more automated method came into use, using computers and automatic random number generators. The idea here was to influence the source of randomness, whether it was a radioactive speck and Geiger counter or a chip with electrical noise, which were supposed to give utterly random streams of numbers. But again here, though the first results by the inventor, German physicist Helmut Schmidt, were well above chance, later experiments with tighter controls were not quite as successful. In general the boredom effect seemed to plague these lab tests: Whenever a new series started up, typically the first few days or weeks would return wonderfully positive results, but after some weeks or months or years this would fade away and the initial positive values would be watered down by lukewarm or negative results. It thus became apparent that psi, if it exists, is fuelled by emotion. This explains why most occurrences of telepathy centre round deaths or disasters. Thus Caesar's wife was said to have dreamt of her husband's death on the night before his assassination. Abraham Lincoln recorded a dream he had just over a week before his assassination. From "Recollections of Abraham Lincoln, 1847-1885" (Ward Hill Lamon, 1911):

'About ten days ago, I retired very late. <snip> I soon began to dream. There seemed to be a death-like stillness about me. Then I heard subdued sobs, as if a number of people were weeping.

There I met with a sickening surprise. Before me was a catafalque, on which rested a corpse wrapped in funeral vestments. Around it were stationed soldiers who were acting as guards; and there was a throng of people, some gazing mournfully upon the corpse, whose face was covered, others weeping pitifully. 'Who is dead in the White House?' I demanded of one of the soldiers 'The President' was his answer; 'he was killed by an assassin!' Then came a loud burst of grief from the crowd, which awoke me from my dream.'

After Lincoln's murder (by assassin!) his casket was, in fact, put on a platform in the East room where soldiers were stationed to act as guards. Still, one could argue that he had often grounds to fear an assassin.

There are many more such stories. Some more recent ones are described in Guy Lyon Playfair's book on twin telepathy (Playfair, 2002). For example, there was the case of the McWhirter twins, Norris and Ross, memory champions central in producing the Guinness book of records for many years. Ross was murdered by the IRA in 1975. At just that time, Norris and his family were preparing to go to his daughter's school play. Alasdair, his son, described to Playfair how he was waiting in the drawing room with his father when Norris suddenly slumped back in his chair, dazed. When he had recovered a few minutes later from his shock, the police rang with the bad news.

There are many such stories of twins knowing when something serious happened to their twin by no apparent means other than maybe 'telepathy'. And empathy between the twins cannot explain such cases when injuries or death are felt across distances of many (sometimes hundreds or thousands) kilometers. That there is such a strong link between twins has often been mentioned. Note that in the old days before the Human Genome project revealed how few genes we actually have, genetic 'hard-liners' were pointing to cases of identical twins separated at birth and growing up in different countries yet doing exactly the same sort of thing in almost absurd detail. The explanation proffered then was that the same genes gave same behavior. But with the lower gene count this became untenable, as there were scarcely enough genes to sketch even the most rudimentary of brain characteristics. Thus the documentary films dwelling on this 'proof' of genetic determinism were quietly withdrawn from circulation. Could a better explanation be that the telepathic link kept both twins 'in tune' with each other?

One author who made this mistake was Steven Pinker, who in his book 'How the Mind Works' lays great emphasis on the synchronization between separated identical twins. That book was published in 1998, however, some years before the number of genes in the genome was announced. If he had been writing it now, this aspect would, I suspect, have quietly been removed.

Consider for example the case of the 'Jim twins', described by Segal (2002):

* As youngsters, each Jim had a dog named "Toy."

* Each Jim had been married two times -- the first wives were both called "Linda"

and the second wives were both called "Betty."

* One Jim had named his son "James Allan" and the other Jim had named his son

"James Alan."

* Each twin had driven his light-blue Chevrolet to Pas Grille beach in Florida for

family vacations.

* Both Jims smoked Salem cigarettes and drank Miller Lite beer.

* Both Jims had at one time held part-time posts as sheriffs.

* Both were fingernail biters and suffered from migraine headaches.

* Each Jim enjoyed leaving love notes to his wife throughout the house.

There are just a few such cases. But there are also many cases where identical twins' life patterns rapidly diverge as environment and maybe free will play a greater role. For example in some cases even when not separated one twin is gay and the other straight, one left-wing and the other right-wing politically, etc. This pattern suggests that there is indeed no gene or gene combination responsible for naming your son 'James Allan'. Thus twins may only very rarely resemble each other as uncannily as the 'Jim twins'. This in turn suggests the hypothesis that only in some rare cases is the telepathic or 'non-local mind' link so strong as to give such good agreement as for the Jims, since the probability of all of the above factors being in common purely by random means is ridiculously small. To estimate how small, one would have to get the chance of each of the above factors and multiply them together. Let's say that the beer and cigarette factors were relatively common, while others were less so – e.g. having a dog named 'Toy' or marrying women with the same name twice. As a ball park feeling, a very conservative estimate of the average probability could be about 1 in 30 or 0.0333. To get the combined probability, we multiply the 8 factors to get $P = (0.0333)^8 = 1.52 \times 10^{-12}$, or there was a 1 in 658,000,000,000 chance of all these coincidences simultaneously. I know that statistics purists would argue the details of this estimate, but in the end their corrected estimate would not be so different. This means that you would have to have 660 billion identical twin pairs separated at birth to get the result by pure chance. Unless the entire population of Earth was twins since homo sapiens evolved, this would be essentially impossible. I.e. it is indeed astronomically unlikely by purely random means. So telepathy rears its head again.

I know that many will object to the probability calculation above, saying there might be other unknown biasing factors not mentioned. But then they should complain to Pinker and the others who had seen the 'coincidences' as proof positive that genetic determinism was vindicated. When the missing genes falsified the latter idea, the question was immediately obvious – 'if not genes, then maybe psi?'

Some of the laboratory psi studies were in fact promising. One of these was the Maimonides dream telepathy project, which often found well above chance results in one person 'guessing' what another had dreamt etc. Another more recent case is Dean Radin and coworkers' experiments with 'presentiment' (see section on Free Will in a previous chapter).

I mentioned above some materialist/reductionist philosophers who dismissed psi out of hand without bothering to read up on research in this area. Some philosophers have, however, considered how important psi could be for the philosophy of mind. One of the philosophers in this field was Charlie Dunbar Broad (1887 – 1971) (see e.g. http://en.wikipedia.org/wiki/C._D._Broad). In the Wikipedia article, his points on psi and philosophy ovelap with mine to a large extent (I didn't know of him until recently, so it just shows that great minds think alike (via telepathy maybe ☺)):

1. Backward causation, the future affecting the past, is rejected by many philosophers, but would be shown to occur if, for example, people could predict the future.

2. One common argument against dualism, that is the belief that minds are non-physical, and bodies physical, is that physical and non-physical things cannot interact. However, this would be shown to be possible if people can move physical objects by thought (telekinesis).

3. Similarly, philosophers tend to be sceptical about claims that non-physical 'stuff' could interact with anything. This would also be challenged if minds are shown to be able to communicate with each other, as would be the case if mind-reading is possible.

4. Philosophers generally accept that we can only learn about the world through <u>reason</u> and <u>perception</u>. This belief would be challenged if people were able to psychically perceive events in other places.

5. <u>Physicalist</u> philosophers believe that there cannot be persons without bodies. If ghosts were shown to exist, this view would be challenged.[4]

Other Broad-minded (☺) philosophers prepared to consider Psi include William James, who we have already discussed, and several 'Integral' philosophers such as Ken Wilber. Indeed any of the philosophers dealing in the 'Perennial Philosophy' related to ancient Hindu etc. thinking tend to consider Psi as a logical extension of the nature of the mind. In Indian or Chinese philosophy, or modern Integral or Transpersonal philosophy, meditation is said to lead to higher levels of consciousness. Some models of the mind only go as far as Freud's three levels sub-conscious, conscious, superego. Others add layers above this, corresponding to the ancient Indian 'subtle levels'. These levels include spirit and soul and can lead to 'special powers': levitating Tibetan monks, The great saint Milarepa who could withstand icy cold in a thin robe, Tai Chi or Qi Gong masters being able to knock down opponents with 'energy' outbursts, etc. This again leads onto the idea of the 'subtle body', which Wikipedia describes as follows:

"According to various esoteric, occult, and mystical teachings, living beings are constituted of a series of psycho-spiritual **subtle bodies**, each corresponding to a subtle plane of existence, in a hierarchy or great chain of being that culminates in the physical form.

It is known in different spiritual traditions; "the resurrection body" and "the glorified body" in Christianity, "the most sacred body" (*wujud al-aqdas*) and "supracelestial body" (*jism asli haqiqi*) in Sufism, "the diamond body" in Taoism and Vajrayana, "the light body" or "rainbow body" in Tibetan Buddhism, "the body of bliss" in Kriya Yoga, and "the immortal body" (*soma athanaton*) in Hermeticism.[1] The various attributes of the subtle body are frequently described in terms of often obscure symbolism: Tantra features references to the sun and moon as well as various Indian rivers and deities, while Taoist alchemy speaks of cauldrons and cinnabar fields."

So there is a rich tradition down all the ages of humankind of taking the psychic dimension seriously. There is also a lot of modern evidence for extraordinary experiences, rare but consistent in their characteristics. This all points to the strong possibility that 'psi' is a real phenomenon. It is a scandal that this important area of human experience has been ignored by mainstream science simply because it does not accord with the dominant world-view of materialist reductionism. It would be nice to think that the 21st century could begin again with serious investigation of the paranormal. But it doesn't seem likely and so the 19th century will continue to shine as the golden age of psychic research.

A modern philosopher who has risked professional approbation in his investigations of the paranormal is Stephen Braude (another synchronicity that would appeal to Braude, who investigated meaningful coincidences, as the 2 main psi-philosophers, Broad and Braude, have similar names). In his book "The Gold Leaf Lady" (Braude 2007), Braude also indicates the significance of psychic events for the mind-brain debate, if not for the soul-mind-brain debate, as in post-mortem survival (the theme of another book of his (Braude 2003)). Braude is, I think, correct in concentrating on cases of macro-PK, as micro-PK runs the risk of disappearing in the statistical noise, while macro-PK involves large scale events such as floating tables, bending spoons, levitation, materialization, etc. Braude personally investigated several such cases. The following is based on my review of the book on Amazon.com.

'I Already know Braude's special brand of intelligent writing, laced with humour here and there, e.g. from his editorship of the Journal of Scientific Exploration. This book is the most personal I've read from him yet and it really is, as others here say, well written and accessible, as well as being full of interesting ideas. Most of the cases reported have something going for them. I had feared from the introduction that the chapters on the "subject from Hell" and "subject in Hell" would be not so riveting. But they are, like all else here. Especially heart-rending was the story of

Dennis Lee, shabbily treated and yet willing to cooperate to the utmost to prove himself. The tragedy in his case was that he, in contrast to the Subject from hell, did seem to have a psychokinetic talent. But though it was apparently demonstrated outside the lab, he had no luck in the lab, partly due to his first harrowing experience in New York. I think his case could be made more forcefully posthumously (he died a tragic death), though Braude has done a good enough job here. The Serios chapter is good as well, and his case is not nearly as tragic as Lee's, in that he did get to test well and a true mystery appears to have been demonstrated in strictly controlled experiments: even where he didn't use his 'gizmo' or was too far away for that. Randi is exposed for his dishonest dealings in reneging on his promise to reproduce the results under similarly strict conditions, as are closed-minded 'sceptics' in general (e.g in the historical cases of Home and Paladino). The eponymous lady is also very interesting and one hopes that some more video etc. evidence will be gathered, as she is, unlike most of those discussed, still alive. Hers is a fascinating gift, the seeming materialisation of shining metal on her skin (hundreds of witnesses) - and she also seems to have ESP gifts such as locating crime victims and predicting crimes to come. Another predictor is the author's own wife, whose astrological successes might be an example of macro PK on a large scale or an ESP-mediated interpretation of the stars. Some of the ideas in that direction remind of the book "The PK man" - another thought provoking case. She appears to be quite a lady - developed her own technique for astrological readings that seemed so successful that football teams valued her advice and came further when they heeded her detailed predictions. It's the most convincing case I've yet heard for astrology. So, 5 stars to this readable and enjoyable book. Even the rants are not so bilious, though I could feel my blood boil at the treatment meted out to Prof. Braude by closed-minded colleagues over the years. Maybe he should emphasize that there are wonderful counter-examples in some very helpful colleagues and friends.'

So – all in all there is quite a bit of evidence for the paranormal. As Braude also says, it's no use dismissing all witness evidence as episodic or casting doubt on the quality of witness reports, as all court cases are based mainly on witnesses' evidence, and the case for many of these paranormal occurrences is actually tighter than those that have locked some criminals away for years. For instance, for the 'Gold Leaf lady', Katie, herself, there were many witnesses to metallic foil appearing on her face or arms or legs. The film evidence caught some of this, though only once in the initial moments of appearance, and then not under ideal circumstances: she has no conscious control over it and can't summon it to order. The foil is also being examined: university labs analyzed it and found it similar in composition to commercial foil – though differing in concentration of some elements from the commonest brands. If more money were available for the case then a more detailed forensic check should therefore be able to indicate where the foil was produced, in the case of it being a fake or having 'teleported'.

Failure to find any manufacturer using this concentration of elements would rule out obvious fraud. But again, the many respectable witnesses imply there was more to it than simple fraud.

If a well paid government study were to be made of such macro-PK cases, it would be interesting to see how much data they gathered. Probably a huge budget would be needed, as the number of people who report anomalous events is staggering. When I think of my 2 neighbours, just 3 and 4 doors away, one with his self-bent-fork (during a Geller show on TV) and the other who saw a disk with coruscating colour patterns fly over our group of houses in the early 1990s, then the number of unreported cases must be in the many millions in each large country. And these 2 neighbours are extremely conservative businessmen – one a 'pillar of the community', so they never would have reported this except in private conversation.

So on the one hand we have the brain research which implies more and more that elements of intelligence and personality depend on areas of the brain. On the other we have so much evidence for paranormal influence of the mind beyond the brain. A materialist would argue that the paranormal is all nonsense (without examining the evidence) and the equation 'brain = mind' is valid. An idealist could argue that brains are illusions and just symbolic of states of mind. Dualists lie somewhere in between. Then brain damage, according to the latter, could be a sign of a decay of connection of the dualist mind to the world of actions on what seems to be the Earth. The jury, I believe, is still very much out on the ultimate nature of things.

Conclusions: The Meaning of life, the Universe & Everything

Several books in recent years have tackled the title of this chapter. One was Baggini's – and I think I presented enough samples of work in the field of parapsychology in the last chapter to show that there is a rich tradition there which Baggini was ill-advised to dismiss out of hand without even glancing over what has been written in the field, on the grounds that it was probably pointless, like everything in his philosophy. This may be a lesson that a key to the meaning of life might be lurking round the next corner: always maintain your curiosity and openness to new ideas, however abhorrent they may seem at first. For instance, if you could prove the reality of psychokinesis, this would invalidate the strict brain = mind = computer equivalence of the reductionist-materialist camp.

I agree with Michio Kaku in his book 'Parallel Worlds' that physicists are the one type of scientists who can say the word 'God' without flinching and I must say I enjoyed the book thoroughly. But Michio may be in for a surprise if Extended Heim Theory (see the first chapter above) is right about the anti- gravity space drive, as in his well written speculation of how civilisation types of increasing energy and knowledge should take centuries or longer to begin first trips to the stars, through laborious wormholes or via Alcubierre 'warp' drives (also requiring massive amounts of enegy), he makes two assumptions which in a few years may be shown no longer to hold: First, we may not need to plod along with ion drives to Mars for the next century, if the Heim-drive works. Even on the energy front he may have been pessimistic as that other spinning magnet system, Steorn, may finally present a decent prototype to the world of their magnet motor this year. This may be a lot of maybes, but they are actually not too unlikely as I know from personal experience (I had a personal demo of Steorn's free energy in April 2008 and was at the 'final proof of Over-unity' in January 2010 and I am in touch with leaders in the field of Heim Theory and anti-gravity).

Another thinker with a take on a theory of everything including mind / consciousness is Ervin Lazlo. In his book 'Science and the Akashic Field' (Lazlo, 2004), he makes a case for the quantum vacuum as a potential storage area for all that has passed since the Big Bang, and a potential place for an afterlife for consciousness after death of the physical body. While an interesting proposal, I admit that if there is a TOE that can make an afterlife plausible then it would be Heim Theory, as th extra dimensions are ascribed qualities such as meaning, and Heim himself saw the possibility of ' immortality in the 6th dimension'. This is the more speculative aspect of his theory, just as the Akashic field is a very speculative aspect of the quantum vacuum.

So, if one wants to consider the meaning of life, the universe and everything, one may content oneself with the world of the physicalist: atoms and energy randomly spewed out from a chaotic quantum explosion with no rhyme or reason and no plan: a blind watchmaker fumbling in eternal darkness. If this image is not so appealing, one may turn to the ancient traditions which imply something more in Heaven and Earth than are dreamt of in the philosophy of the materialists.

We have seen, in my brief survey of the philosophy of mind, that 1st person subjective aspects are at least as important as 3rd person objective studies. We are more than just emergent phenomena based on brain processes – subjectivity may extend right through the animal kingdom, breathing meaning into the pointlessness of the objective. The genetic stock-taking showed that it may be that more than the genome is needed to explain bodily and mental complexity: could some of the information be stored not only epigenetically, but epi-bodily, as in morphic fields or souls? Paranormal studies imply that there may indeed be life beyond the brain, and not all is the atoms of Democritus and Newton, but a more interconnected, holistic, quantum or even paranormal / numinous scheme is the reality.

References

Baars, B. (1997), 'In the Theater of Consciousness', Oxford University Press. ISBN 0195147030

Barbour, J., 1999: 'The End of Time', Weidenfeld & Nicholson

Baggini, J. 2005, 'What's it all about?' Granta Books, London.

Braude, S. 2003, 'Immortal Remains' Rowman & Littlefield, ISBN-13; 0742514720.

Braude, S. 2007, 'The Gold Leaf Lady' Univ. of Chicago Press, Chicago, ISBN-13; 978-0-226-07153-7.

Brooks, M. 2008, '13 Things that don't make Sense' Doubleday, Random House, New York, ISBN 978-0-385-52068-3.

Capra, F., 1996, 'The Web of Life' – Anchor Books, Random House, New York.

Chalmers , D., 1995: 'The Puzzle of Conscious Experience', Scientific of, American, December 1995.

Chalmers, D.(1996) 'The Conscious Mind', Oxford University Press,, ISBN 0-19-511789-1 .

Chalmers, D.,1994: 'Facing Up to the Problem of Consciousness', Journal of Consciousness Studies 2(3):200 - 19, 1995

Choi, C., 2003, 'Scaled up Superposition', , Scientific American, February 2003

Conway Morris, S., 2003, 'Life's Solution: Inevitable Humans in a Lonely Universe' , Cambridge University Press (Sept 2003)

Dahlbom, B. (1993) 'Dennett and his Critics', Blackwell, ISBN 0-631-19678-1.

Dawkins, R. (1989) 'The Selfish Gene' p 197 - 198 Oxford University Press, UK

Dennett, D.C. (1995) 'Darwin's Dangerous Idea' p 515 - 517, Penguin, UKIt

Dennett, D.C. (1993) 'Consciousness Explained', Penguin, ISBN 0140128670

De Quincey, C. (2002) in 'Cognitive Models and Spiritual Maps' (pp. 177 – 208), Imprint Academic, ISBN 0-907945-13-4.

Eisler, R., 1987, 'The Chalice and the Blade', HarperSanFrancisco ISBN: 0-06-250289-1

Gribbin, J. 1995: 'Schroedinger's Kittens and the Search for Reality', Weidenfeld & Nicolson, London, 1995.

Hawking, S., 1996: 'A Brief History of Time', Bantam

Heidegger, M., 1997: 'Being and Time', State Univ of New York Pr; (October 1997)

Horgan, J. 1996: 'Schroedinger's Cation', Scientific American, June 17, 1996.

James , W., 1890, 'The Principles of Psychology', New York, Henry Holt and Company

Lazlo, E., 2004, 'Science and the Akashic Field: An Integral Theory of Everything', Inner Traditions International, 2004, ISBN-10: 1594770425, ISBN-13: 978-1594770425.

Libet, 2003: 'Can Conscious Experience affect brain Activity?', Journal of Consciousness Studies 10, nr. 12, pp 24 - 28.

Llinas, R., 2002, "I of the Vortex : From Neurons to Self ", MIT Press (Feb 2002)

Lovelock, J.,1979, Gaia, Oxford University Press, New York.

Nagel, T. 1974. What is it like to be a bat? Philosophical Review 4:435 - 50.

Penrose, R., 1990, 'The Emperor's New Mind: Concerning Computers, Minds and the Laws of Physics', Vintage (Aug 1990)

Penrose, R., 1995. 'Shadows of the Mind', Vintage (Sep 1995)

Pinker, S., 1999. 'How the Mind Works', Penguin Books, U.K., ISBN 978-0-1402-4491-5.

McCrone, J., 1997, 'Wild Minds', New Scientist, 13 December 1997.

Pagán Westphal, S., 2004, 'Wild Minds', New Scientist, 3 June 2004.

Malik, K., 2001: 'Man, Beast and Zombie', Phoenix

Margulis, L. and Sagan, D., 1995, What is Life? – Simon & Schuster, New York.

McGinn, C. 1999, 'The Mysterious Flame – conscious minds in a material world', Basic Books, New York.

Mestel, R. 1993. 'The Mice Without Qualities', Discover Magazine, Vol. 14, nr. 3, March 1993.

Metzinger, T. (1995), "Faster than thought: holism, homogeneity and temporal coding" in Conscious Experience, ed. T. Metzinger (Exeter: Imprint Academic).

Nagel, T. 1974. What is it like to be a bat? Philosophical Review 4:435 - 50.

Playfair, G.L., 2002: "Twin Telepathy.", Vega, ISBN 1-84333-686-3.

Pratchett, T., Stewart, I. and Cohen, J., 2003: "The Science of Discworld II; The Globe", Import books, ISBN 0091888050.

Radin, D., 2006: " Entangled Minds.", Paraview Pocket Books, ISBN-10: 1416516778.

Sacks, O., 2001: "The Man who Mistook His Wife For a Hat.", Picador, ISBN 0330294911.

Searle, J., 1980: "Minds, Brains, and Programs." Behavioral and Brain Sciences 3, 417 - 424.

Segal, N., 2000: " Entwined Lives.", Plume, ISBN 0-452-28057-5.

Seymore, J., and Norwood, D. 1993, 'A Game for Life', New Scientist, 139, nr. 1889, 23 – 6.

Shady, R. Haas, J. Creamer, W. (2001). Dating Caral, a Preceramic Site in the Supe Valley on the Central Coast of Peru. Science. 292:723-726. PMID 11326098 [1]

Sheldrake, R., 1981, A New Science of Life – Tarcher, Los Angeles.

Smolin, L., 2006, "The Trouble with Physics" – Houghton Mifflin, New York, ISBN-13 978-0-618-55105-7 .

Spork , P., 2009: Der Zeite Code', Rowohlt Verlag GmBH Reinbeck bei Hamburg, ISBN-13 978-3-498-06407-5

Stapp , H.P., 1993: 'Mind, Matter, and Quantum Mechanics', Springer Verlag Berlin, Heidelberg

Sutherland , K., 1997, jcs-online thread: http://www.imprint.co.uk/online/homuphob.html.

Targ, R. and Puthoff, H. 1974. Information Transmission Under Conditions of Sensory Shielding. *Nature* 251: 602-604

Von Lucadou, W., 1995, "Psi-Phänomene – Neue Ergebnisse der Psychokinese-Forschung", Insel, FF am Main & Leipzig. ISBN 3-458-33809-8.

Weinberg, S., 1993, 'The First Three Minutes: A Modern View of the Origin of the Universe' (1977, updated with new afterword in 1993, ISBN 0-465-02437-8).

Whitehead, A.N., 1978: 'Process and Reality', Free Press, ISBN 0-02-934570-7

Winston, R., 2002: 'Human Instinct – How our primeval impulses shape our modern lives', Bantam Books, ISBN 0-553-81492-3

www.ingramcontent.com/pod-product-compliance
Lightning Source LLC
Chambersburg PA
CBHW030930180526
45163CB00002B/520